WHAT EVERY ENGINEER SHOULD KNOW ABOUT
PRACTICAL CAD/CAM APPLICATIONS

WHAT EVERY ENGINEER SHOULD KNOW
A Series

Editor

William H. Middendorf

Department of Electrical and Computer Engineering
University of Cincinnati
Cincinnati, Ohio

Other volumes in preparation

WHAT EVERY ENGINEER SHOULD KNOW ABOUT
PRACTICAL CAD/CAM APPLICATIONS

John Stark

Coopers & Lybrand Associates
London, England
Geneva, Switzerland

MARCEL DEKKER, INC.　　　　　　New York and Basel

Library of Congress Cataloging-in-Publication Data

Stark, John, [date]
What every engineer should know about practical
CAD/CAM applications.

(What every engineer should know ; 17)
Bibliography: p.
Includes index.
1. CAD/CAM systems. 2. Computer-aided design.
3. Engineering design--Data processing. I. Title.
II. Series: What every engineer should know ; v. 17.
TA174.S768   1986   670'.285   86-13412
ISBN 0-8247-7593-7

MARCEL DEKKER, INC.
270 Madison Avenue, New York, New York 10016

Current printing (last digit):
10   9   8   7   6   5   4   3   2   1

PRINTED IN THE UNITED STATES OF AMERICA

D
620.0042
STA

*For Alicja
and
for Jasna,
"une fille formidable"*

PREFACE

The aim of this book is to describe the necessary background for a successful implementation of CAD/CAM in the mechanical engineering environment. Although CAD/CAM means different things to different people, there is general agreement that it offers the possibility to increase productivity and for this reason is of strategic importance. In this book, CAD/CAM is regarded as a tool for improving the quality, value, and flow of information in design engineering (e.g., in design, drafting) and in manufacturing engineering (e.g., in NC programming, process planning). Such improvement leads to reduced cycle times, reduced costs, and improved quality. However, productivity gains do not follow automatically from the use of CAD/CAM. To be successful, CAD/CAM must be properly implemented and managed.

CAD/CAM is just one of many computer-based tools used throughout a manufacturing company. In the past, companies have tended to computerize isolated parts of the manufacturing process. These parts are often referred to as "islands of automation." Examples of islands are CAD/CAM, MRP (manufacturing resource planning), SFC (shop floor control), factory automation (e.g., numerically controlled machine tools), and office automation (including word processing).

As a result of increasing power and falling cost of computer hardware and software, it is now apparent that before long all operations within a manufacturing company will become computer assisted. It will be necessary to ensure that information and commands can flow smoothly between islands of automation. For example, information in a parts list generated in a CAD/CAM system will be needed by the MRP "island of automation" to calculate material requirements and to schedule machine loadings. A machine tool control program generated in the CAD/CAM system will have to be transferred to the shop floor to control the machine. If the machine breaks down while the part is being machined, and will be unavailable for three months, then that information will have to be transferred back to the MRP island so that machines can be rescheduled. The term *computer-integrated manufacturing* (CIM) is often used to describe such a manufacturing entity in which all decisions, commands and information flows rely on assistance from computer systems. In CIM, the various business functions whether they be engineering, production, sales, or finance will all be computer aided, and information flows between them will be computer based. Whereas computerization of an "island of automation" such as CAD/CAM leads to a local increase in productivity, implementation of CIM will lead to a global increase in productivity.

Companies need to have an overall CIM implementation policy, under which individual functions can be computerized as efficiently as possible. This book describes how the design engineering and manufacturing engineering functions may be successfully computerized (i.e., how CAD/CAM may be successfully implemented). CAD/CAM is concerned with generating and using information that has "belonged" in the past to different departments. The information it generates will be used within various departments. For CAD/CAM to be really productive, information must be seen as a global, company-wide resource, not as a collection of fragments of data each of which is the jealously guarded property of a particular department. For this to be possible, company structures must change to reflect the change in the way in which information is treated. Only top management can cause the necessary changes to occur.

Manufacturing companies' organizations have changed in the past. Before the Industrial Revolution, manufacturing was essentially carried out by craftsmen producing small volumes of customized products. The Industrial Revolution, which started in the middle of the eighteenth century, was technology based. The invention and application of new machines and technologies led to productivity increases. At the end of the nineteenth century, productivity was further increased by the introduction of high-volume production of standard parts. In the early twentieth century, another opportunity arose to further increase productivity. New management-oriented organizations were set up in which the overall functions of a manufacturing company were split up into specialized tasks, each of which could be carried out by one worker. A foreman supervised a group of workers. Middle management supervised foremen, and so on. The technique worked well when companies concentrated on high-volume production of standard products that only rarely underwent significant change. In the 1980s, however, the need is to produce customized products in small volumes for a fast-changing consumer-oriented market.

In management-oriented organizations, tasks were specialized and an individual no longer had a complete view of the overall manufacturing process. One could be a "design engineer" or a "manufacturing engineer," or a "drawing office manager" but not just an "engineer." The information, knowledge, and ability to manufacture a product was divided up between different departments. The departments became empires and surrounded themselves with walls. As a result, information flow was impeded. In the worst case, "design engineers" designed products that could not be manufactured. It became impossible to respond quickly to fast-changing markets. The walls had to come down.

It is not clear whether change in organizational structure results from increased use of computer-based techniques, or whether computer-based techniques have to be implemented to allow organizational changes to take place to meet the challenges of new markets. It is clear though that organization changes allowing free flow of information will occur. One of the most important techniques for improving the quality, value, and flow of computer-based information is CAD/CAM.

Many attempts to implement CAD/CAM in industry have not led to the results hoped for. Some companies have bought a CAD/CAM system and then found that they did not know how to use it, or that it was not suitable for their particular type of product. Others have bought a system, and then found it necessary to buy a replacement or complementary system. Even among companies that claim to have successfully introduced CAD/CAM, some say that although they could not go back to handling their applications by manual methods, there is no directly measurable productivity gain. Almost all companies find that the productivity gain varies significantly from one application to another. There are some applications for which a loss in productivity may result from the use of CAD/CAM.

Many users would say that CAD/CAM has not been successful because CAD/CAM systems are not all that they claim to be. There is a certain amount of truth in this since many of the systems do not really aid the design process—only some of the downstream processes such as drafting. Consequently, there is a limited productivity gain. In the future, as CAD/CAM systems are improved, it will be possible to use them in the complete design engineering and manufacturing engineering cycle, and increased productivity gains will result. However, even the best CAD/CAM system will not flourish under poor management. Due to rapid technological advances, managers today have less understanding of the processes that they try to manage than at any time in history. Many of them received their formal education before the computer was invented and very few of them have been trained to manage computer-assisted techniques.

There is a tremendous need for both managers and users to learn what CAD/CAM really is and how it can be successfully implemented. Part I of the book gives basic information on CAD/CAM and describes how to select, implement, and run a CAD/CAM system. The first chapter in Part II describes the overall state of CAD/CAM today in different industrial sectors and for different manufacturing technologies. It is based on visits to companies and discussions with CAD/CAM users and managers in the United States, Europe, and Asia. It shows that the question companies should be asking themselves is no longer "Should we use CAD/CAM?" but "Why are we not using CAD/CAM?" or "Are we using CAD/CAM as effectively as possible?" The second chapter of Part II contains case studies of the selection, implementation and use of CAD/CAM by a variety of companies. The last chapter of Part II is based on actual implementation and use of CAD/CAM in many companies. These companies, of various sizes, in different countries, in different sectors, and at different stages of use of CAD/CAM, represent a wide spectrum of mechanical engineering CAD/CAM usage. From their experience it has been possible to identify ingredients required for successful implementation of CAD/CAM. This will prove to be helpful not only to those starting to use CAD/CAM but also to those wanting to improve their existing installation.

John Stark

CONTENTS

Part I

Selection and Implementation

1

INTRODUCTION TO CAD/CAM

Computer-aided design/computer-aided manufacturing (CAD/CAM) is a computer-aided technique for improving the efficiency of design engineering and manufacturing engineering activities. A CAD/CAM system is the tool used for applying the technique. Some of the activities (or application areas) are shown in Table 1.1. There are two apparently conflicting ways of looking at CAD/CAM. One of them sees it as a tool for improving the productivity of a particular application such as numerical control (NC) programming or drafting. The other viewpoint sees it as a way of treating and transferring information in a more efficient, secure,

Table 1.1   Application Areas Assisted by CAD/CAM

Design engineering
  Preparation of quotations
  Conceptual design
  Styling
  Finite-element analysis
  Simulation
  Kinematics
  Engineering design
  Detailed design
  Schematics and wiring diagrams
  Quality assurance
  Drafting
  Parts lists
  Technical publications

Manufacturing engineering
  Tool and fixture design
  NC machine tool programming
  Robot programming
  Quality control machine programming
  Process planning
  Preparation for automated testing
  Plant layout
  Material handling simulation
  PLC programming

and useful way than is possible with manual methods. The two viewpoints are not mutually exclusive. One of the problems of CAD/CAM implementation is to combine the requirements for the more easily quantifiable "productivity" viewpoint with the less tangible but often greater benefits that can be derived from improving the quality of information flow.

When considering the overall engineering and production activities of a company it is clear that the use of computers is not limited to CAD/CAM but extends into other areas such as manufacturing planning and control systems, management information systems, structural analysis, robotics, purchasing, testing, marketing, and so on. Computers have been used in these areas for different lengths of time. Often the first activities in a company to be assisted by computer were those in the financial sector. However in some companies, computers have been used since the 1950s both in design engineering, for example in finite-element analysis (FEA), and in manufacturing engineering, for example in programming numerically controlled (NC) machine tools.

1.1 BATCH PROCESSING

In the early days of computing, applications were traditionally carried out in "batch," that is, a complete task (or job) was first defined by the user and submitted to the computer. The computer processed the complete job without further interaction with the user, and then produced output. Several hours could elapse between the submission of the job and receipt of the output. The user would then take the output, generally a listing, away and study it at the desk. As a function of the results, the input parameters of the job could be modified and the whole batch process repeated. In the batch process, the computer is mainly used as a rapid calculator to solve a particular problem faster than would be possible by manual techniques. Although the calculation process is rapid, the overall process is slow (the user may have to wait several hours for results if the computer is busy running other jobs) and the information transfer techniques lengthy. Typically the user would have entered all input data at a keyboard, and output results would have been transferred by listing to other areas of the overall design engineering and manufacturing engineering process. In these areas, data would be read off the listing and entered into further manual or batch calculations.

1.2 INTERACTIVE COMPUTER GRAPHICS AND CAD/CAM

Two features that distinguish CAD/CAM from traditional engineering computing are its use of interactive graphics techniques rather than batch techniques, and its potential to reuse information (e.g., part data) in the computer both within an application area and across the boundaries of different application areas. The requirement to reuse part data leads to two further distinguishing features. The first of these is the need to model (in the computer) the geometry of the part in such a way that sufficient information is available for reuse later both within the same application and within other applications. The other feature is the need to be able to store, transmit, and retrieve part data. As an example of the reuse of part data, consider the case of turbine blade manufacture. First of all, the geometry of the turbine blade has to be designed. While using the CAD/CAM system for design, a model of the geometry of the blade is built up. Once that model exists it can be made available both to engineers carrying out stress analysis and to those preparing the NC program. Savings will occur because the geometry does not have to be recreated for analysis and NC tool path generation, and also because all three application areas are using exactly the same geometry data, thus reducing the risk of errors. These types of savings can be translated into direct benefits such as a reduction in time cycles, a reduction in costs and an improvement in product quality. However, these direct benefits are just an expression of the fact that information, perhaps a company's most important asset, is being used more efficiently.

Computer-based interactive graphics techniques involve use of a computer and a television-like screen on which drawings (graphics) and results (text) can be displayed. The user interacts with the computer via an input device such as a mouse or a light pen. The computer carries out the necessary calculations, controls the image on the screen in real time, and can output results in the form of drawings, tapes, listings, etc. In many cases, less than one second will elapse between the time at which the user inputs a command and the time at which the user sees the result at the screen. This almost instantaneous process gives the user the impression of carrying out an interactive dialogue with the system. The visual nature of interactive graphics increases the user's understanding of a task and reduces the potential for some of the errors and oversights associated with manual procedures. As well as calculating faster than humans, the computer can carry out repetitive and uninteresting work more efficiently than humans. It can also store large quantities of information and retrieve required information very quickly. The computer and interactive graphics can aid the user to substantially increase quality. Many engineering tasks are still run in batch because they require a lot of computer processing time. It would be inefficient and wasteful for a user to sit in front of an empty screen for several minutes waiting for the result of a complex calculation. However, even those tasks that have a high batch content can often be aided by the use of interactive graphics. As an example, consider the case of finite-element analysis. Although the analysis calculations may take several hours and be run in batch, the preparation of the finite-element mesh (the pattern of small elements into which the part is decomposed) can be carried out in an interactive fashion with the help of a CAD/CAM system. Understanding of the results of the calculation can be increased by the use of graphics techniques. Similarly the generation of an NC tool path can be eased by using a CAD/CAM system. Although the actual calculation of the tool path may be carried out in batch, it is very useful to display the path and to simulate the movement of the tool at the screen.

1.3 SOME APPLICATIONS OF CAD/CAM

Apart from finite-element analysis and NC programming, other applications can benefit from the use of CAD/CAM. Engineering drawings can be built up (or "drawn") on the screen, stored by the computer, and afterwards recalled to the screen and, if necessary, modified. Once the drawing is satisfactory it can be automatically drawn by a plotter. Another application that benefits from the use of CAD/CAM is simulation. The designer may investigate on the screen, and from a variety of viewpoints, how a mechanism moves and whether it collides with another part. As well as aiding the programming of NC machine tools such as lathes and milling machines, CAD/CAM can be used when programming robots and quality control machines. In the manufacturing engineering area it can also aid the design of tools and the preparation of process plans. CAD/CAM can be seen to be a very useful computer-based technique using interactive graphics techniques. It can be used for many applications throughout design engineering and manufacturing engineering. However, it must not be forgotten that the CAD/CAM system is only a tool. It does not design or analyze or manufacture. These tasks continue to be carried out by people (e.g., designers), programs (e.g., finite-element analysis), and machines. There is, though, one new "application" which has come into existence because of CAD/CAM. It is called geometry modelling.

1.4 GEOMETRY MODELLING

It has been seen that among the distinguishing features of CAD/CAM are its reuse of part data and the consequent need for a computer-based model of the part that can be used in several application areas. In the past, with manual techniques, information on part data was transmitted from person to person on drawings. Manually produced drawings of typical mechanical parts often do not exactly reflect what the part is—they tend to be incomplete, ambiguous, and incorrect. Computer systems do not currently have enough intelligence to decide what the person who produced the drawing was really trying to describe. Even if the drawing does give a meaningful description, it may not contain the information required by another application area.

The term "product modelling" covers the process of building up a computer-based model containing all the necessary information on the part or product. This information includes attributes such as geometry, color, material and so on.

Geometry modelling is the process of building a model (in the computer) that contains all the necessary information on the part's geometry. The model should be unique (i.e., the part will not be mistaken for another) and complete (it contains all the geometry information required in the various application areas). It will be seen later that there are several different methods for modelling geometry. Early attempts at developing CAD/CAM did not always model geometry suitably. Part geometry was not sufficiently described and systems tended to be specific to one application. The variety of terms used to describe these systems [CAD, CADD (computer-aided design drawing), CAD/CAM, CAM, CAE (computer-aided engineering), etc.] gives rise to misunderstandings.

1.5 PROBLEMS WITH ACRONYMS

The term CAD was originally used to mean computer-aided design and is still sometimes used in this sense in that it is the use of the computer in the conceptual design/engineering design part of the process and includes analysis and simulation rather than drafting. The term CAD has also nevertheless been used to mean computer-aided drafting. CAM was originally used to mean computer-aided manufacture and could be applied equally well to programming an NC machine tool as to scheduling use of the tool or as to the use of the tool to manufacture a part. The term "NC programming" is not ambiguous but does not refer to the entire manufacturing engineering activity. The terms CAP (computer-aided production) and CAPP (computer-aided production or process planning) only seem to confuse the issue.

In this book, CAD/CAM is defined as a computer-based technique to aid design engineering and manufacturing engineering activities. Thus CAD/CAM is computer-aided design engineering and computer-aided manufacturing engineering. However, this definition of CAD/CAM does not englobe all computer techniques in these areas. For example finite-element analysis is not part of CAD/CAM, although, as has been shown, CAD/CAM can improve productivity in this particular area.

This book will consider the computerized manufacturing plant as being made up of four major "islands of automation":

- CAD. Computer-aided design engineering (defining what the product is).
- CAM. Computer-aided manufacturing engineering (defining how to make the product).

- MRP. Manufacturing resource planning (defining when to make the product).
- FA. Factory automation (making the product).

Two other acronyms that describe important concepts but do not currently have agreed meanings are CAE and CIM. In this book they are defined as:

- CAE. Computer-aided engineering. *All* computer-based techniques used in the design engineering and manufacturing engineering areas (i.e., not only CAD/CAM but also analysis, simulation, etc.).
- CIM. Computer-integrated manufacturing. Integration of all computer-based techniques applied to all functions throughout a manufacturing company. These techniques include CAE, MRP, FA, office automation, etc. The "integration" in CIM implies that all decisions and command and information flows in the company are computer assisted.

1.6 CAD/CAM, COMPUTER-AIDED DRAFTING AND NC PROGRAMMING

Is computer-aided drafting, i.e., the use of a computer-based system for producing drawings, an application area of CAD/CAM? Under the above definition of CAD/CAM, the answer is that sometimes it is and sometimes it is not. If the part's geometry has been modelled and stored in the data base, and can be used again in manufacturing engineering, then drafting of the part is an application area of CAD/CAM. On the other hand, if there is no part model in the data base and someone is going to use the drawing to manually input geometry data into the manufacturing program, then drafting is not an application area of CAD/CAM—it is just computer-aided drafting. The distinction is very important, because the savings and increases in productivity and quality that result from use of CAD/CAM occur when the modelled geometry of the part is used all the way through design and manufacturing engineering.

Dissatisfaction with CAD/CAM occurs in many mechanical engineering companies where CAD/CAM has been equated to computer-aided drafting. Although computer-aided drafting can rapidly give productivity gains, these gains are limited. The CAD/CAM screen is not just "an electronic drawing-board." When it is used in the true sense of CAD/CAM it is also "an electronic design medium," "an electronic model maker," and an entry point into a sophisticated information system.

The question also arises of whether NC programming is an application area of CAD/CAM. In conventional NC programming, a part programmer receives a manufacturing drawing of the part on a physical medium and develops the control program either manually or with assistance from a remote or local computer. Part programming carried out in this way requires the manual entry of the geometry of the part and the tool path. Even if interactive graphics techniques are used, this would not be considered an application of CAD/CAM, unless the part programmer uses existing information in the computer, the part geometry. Due to the reuse of information there will be a reduction in data redundancy, elimination of errors in interpretation, and a reduction in cycle time.

1.7 THE IMPORTANCE OF CAD/CAM

Improvement in productivity is welcome at any stage of the manufacturing process. CAD/CAM is important not only because it offers productivity gains at many stages, but especially because it impacts those stages which are most critical to good, fast product design. Even in the earliest stages of design of a product, CAD/CAM techniques can be used. Marketing staff can use data on existing parts and products to produce an accurate quote quickly. Existing designs can be accessed, and minor modifications may obviate the need for complete redesign. In some industries, CAD/CAM is already used at the earliest stages of conceptual design and/or styling. During engineering design, CAD/CAM may be used in geometry modelling, to assist analysis, in drafting and perhaps even in the preparation for NC machining of a prototype. In manufacturing engineering CAD/CAM can be used to assist in the design of tools, to produce process plans, and in programming NC machine tools and inspection equipment. In the future CAD/CAM will become a key technique in "design for manufacture" and "design for assembly."

Some companies report that without assistance from their CAD/CAM system they would not have been able to prepare a winning quote in time. Others report costs cut in half because CAD/CAM allows them to carry out all applications in-house, and thus remove the need for subcontracting. In some cases, several weeks of design effort have been saved just because it was possible to visualize a complex part on the graphics screen from several different angles. The list of successful CAD/CAM implementations is long, and the range of benefits is wide. However, benefits do not come automatically with purchase of a CAD/CAM system, they only come when CAD/CAM techniques are understood, well implemented, and properly managed.

1.8 CAD/CAM AND CIM

CAD/CAM is just a small part of CIM. Other components of CIM include factory automation, office automation, distribution, MRP, SFC, accounting, payroll, and business simulation. All of these computer systems assist different functions of the company. They all handle information. Not all of them produce or use data that will be held in the CAD/CAM system data base. MRP systems, which use bills of material, inventory, order processing and master scheduling data to determine material requirements, the production plan and machine loadings have a close interface to CAD/CAM. CAD/CAM systems can be used to produce parts lists and then the bill of materials required by the MRP system. Part-programming data produced with the assistance of CAD/CAM must be transferred to the shop floor, and any modifications to programs on the shop floor need to be transferred back to the CAD/CAM system data base. When implementing CAD/CAM, it is important to have a global, or CIM, view of the company, and ensure that data and commands can pass freely between the various systems.

1.9 CAD/CAM AND GROUP TECHNOLOGY

Special techniques such as group technology can be incorporated into the use of CAD/CAM. Group technology is a technique that can be used with or without computer assistance. Its underlying philosophy is that time and cost savings can be made if parts and products can be treated as members of families. First, all parts and products have to be classified into families as a function of their characteristics, such as size of part, shape, ratio of width to length, manufacturing technology used, and so on. A code can then be assigned to each part. Initial savings result from the classification process, when it is often found that many similar or even identical parts can be replaced by just one part. Afterwards, once codes have been assigned, savings occur due to reduced redesign of parts and process plans. Before designing a part for a new product, the designer checks to see if the part already exists in a family. If it does there is no need to redesign it. If it does not, it may be possible to create a slightly modified version of an existing family member, thus reducing design time. Similarly it may be possible to use the process plan of an existing part, or to make minor modifications to an existing process plan.

1.10 THE BENEFITS OF CAD/CAM

The overall aim of CAD/CAM is to improve productivity in design engineering and manufacturing engineering. An improvement in productivity can result in a reduction in cost or cycle, or an improvement in quality. These are the major aims of CAD/CAM and they are achieved by improving the productivity of individual engineering operations and by increasing the productivity of the overall process. Before describing the way in which benefits arise it is useful to point out that there is no universal definition of

CAD/CAM productivity, and that a reduction in cost, a reduction in time cycle, and an improvement in quality are often just different ways of expressing the same improvement in productivity.

Consider a completely fictional case in which, by manual techniques, design and drafting of a part takes two people 15 days each or a total of 30 man-days. Assume that with CAD/CAM, a total of only 15 days is required. There are several ways of handling the saving of 15 man-days:

As a direct cost saving, for example, terminate the employment of one of the people, thus making a saving in salary costs (this is a fairly unusual solution, since it implies stagnation of the company's business, and most companies look to CAD/CAM to promote growth);

As a direct time saving, for example by only spending 15 days on design and drafting of this part, the overall design engineering cycle time will be reduced. Hopefully, the part can therefore be produced and sold earlier than would have been possible without CAD/CAM. Since both people are retained by the company, the 15 days "saved" can be used to design another part, which can also be produced and sold earlier than would otherwise have been possible.

As an improvement in product quality: continue to spend 30 days on design and drafting of the part, but investigate more design alternatives and carry out more thorough checks of the design;

As an increase in volume of business: spending the required 15 days on the part, but spending the 15 days "saved" on preparing a proposal and quotation to compete for a tender that would otherwise have been ignored.

The results of the benefits of CAD/CAM can be used in various ways, not only in design engineering as in the above example, but also in manufacturing engineering and as seen above in the preparation of quotations. In that activity as well, the question will arise of whether to use the benefits to produce more quotations or better quotations. The answer has little to do with CAD/CAM, it is mainly a question of company strategy.

The benefits of CAD/CAM are perceived in different ways at different levels of the company. At the top management level, the overall productivity gain should be recognized to result from an improvement in the flow and quality of information which in turn leads to reductions in cost and time cycles, and improvement in product quality. At this level it is also important to realize that use of CAD/CAM improves the technological image of the company. Since better quality products can be delivered quicker it can also improve relationships with clients. Use of CAD/CAM may be a strategic tool for the company in producing a new type of product. Table 1.2 lists typical benefits of CAD/CAM use.

1.11 SOME PROBLEMS WITH CAD/CAM

It can be seen that a wide range of benefits can stem from the use of CAD/CAM. However it should be stressed that CAD/CAM is only a tool, that it needs to be properly managed, and that today no single system is "the best" for all application areas and all types of product. In its present state, a CAD/CAM system can neither design a new part, nor make a decision as to which of two designs is the better. The designer uses the system as a tool to design a new part, and it is the designer who makes the decisions. The choice and management of CAD/CAM systems are critical issues, since even though CAD/CAM has a high potential for increasing productivity, it also has a high potential for increasing problems. If the wrong system is selected, if the users are not trained properly, if management is not flexible enough to accommodate the changes resulting from its use, then CAD/CAM will not bring significant productivity improvements. It is worthwhile noting and understanding some of the problems that can arise from use of CAD/CAM.

• CAD/CAM systems currently cost a lot of money. Not all companies can afford the initial outlay. Once the system is installed, its cost has to be recouped. This may lead to an unwelcome increase in overhead charges.

Table 1.2   Typical Benefits of CAD/CAM Use

Better, faster, and more accurate quotations
Increased product design quality
Wider use of analysis and simulation techniques
Possibility to study more design alternatives
The potential to design more complex and/or precise parts
Reduced design and drafting cycle time
Improved capabilities for investigating interferences between parts
Easier updating of designs
Reduced requirements for prototypes
Improved techniques for checking assemblies
Improved visual understanding of a part being designed
Easier calculation of geometric properties
Reduced scrap resulting from improved nesting
Facilitated production of parts lists
Improved quality of documentation
Improved drawing quality
Facilitated production of isometric, perspective, and exploded views
Improved documentation for after-sales service
Reduced NC part-programming time
Facilitated part-programming of complex parts
Production of better NC programs due to simulation of tool paths
Improvement in the potential for reusing existing parts and tools
Reduced requirement for subcontracting
Improved receptivity to modifications requested by customers
Reduced material cost resulting from design optimization
Reduced energy costs and machining time resulting from design
    and manufacturing optimization
Reductions in the problems associated with using paper as an
    information medium
Increased use of standard parts
Reduced transcription errors due to use of same data
Possibility to increase enforcement of standardization procedures

---

- Since the system is expensive, there will be pressure to maximize its use. The users (and the unions) may not appreciate two- or three-shift working.
- Since one of the possible benefits of CAD/CAM use is reduction in direct labor costs, staff and unions may oppose the system.
- If the wrong system is chosen for the company's applications, then a decrease in productivity may occur. Alternatively, even if the right system is chosen, problems can occur if the system vendor does not develop it properly, or goes out of business.
- Users do not like to work with (and hence productivity suffers with) a system that frequently breaks down, is poorly documented, has a poor response time, does not correspond to their work methods, etc.
- Unless the large volumes of data that can be generated with a CAD/CAM system are efficiently managed, it will not be possible to achieve all the benefits that should arise from reuse of data.
- Some users will find some systems too difficult to work with, others will find the same systems too boring and frustrating.
- Even if the right system is chosen it can cause problems if sufficient and suitable training is not given or if the system is not properly installed.

The major causes of unproductive use of CAD/CAM are, however, lack of management commitment, foresight, and willingness to carry through the necessary changes required by extensive use of computer-based methods. The organization of a company throughout most of the 20th century has reflected the need to divide activities up into functions that individuals could manage or carry out. However, the computer and its associated flow of information are only rendered ineffective by these barriers, and must be used in a company-wide manner. Management must implement an organization that reflects the most efficient use of today's company resources, which include not only plant, production equipment, and people but also computer hardware and software, data, and knowledge.

2

THE CONTENTS OF A
CAD/CAM SYSTEM

Although CAD/CAM systems are packaged in many different ways, the basic ingredients are always functionally similar. For the purposes of this brief description it will be assumed that the system is based on a single computer. When more than one computer is used, networking hardware and software, and distributed data management software are also needed.

The basic ingredients of the system are in the form of either hardware or software. The hardware will be made up of a central processing unit (CPU) and peripheral devices for data storage, data input, and data output. The various components are shown in Table 2.1. The software will be made up of six levels. First, the

Table 2.1   Hardware Used in CAD/CAM

The CPU

Data Storage
  - Main memory
  - Mass memory

Data Input
  - Alphanumeric keyboard
  - Function keyboard
  - Graphics tablet
  - Picking devices

Data output
  - Graphics terminal
  - Alphanumeric screen
  - Plotters

Communications equipment

---

basic software that makes the computer and its peripherals work. The second level drives the graphics. The third level, the kernel of the CAD/CAM software, includes the user interface, data management, and data exchange software. The fourth level is the geometry modelling software. The fifth level, applications software, links the lower levels to the various application areas such as drafting, analysis, part programming. The first five levels are usually bought from a vendor. The sixth level is made up of user-written, company-specific software. The six levels of software are illustrated in Table 2.2.

2.1   HARDWARE

2.1.1   Central Processing Unit (CPU)

At the heart of the computer is the CPU which is made up of an arithmetic and logic unit, and a control and command unit. Programs and data are retrieved from the main memory. The CPU

Table 2.2   The Six Levels of CAD/CAM Software

Level 6	User-developed software
Level 5	Applications software (e.g., drafting, NC programming, kinematics)
Level 4	Product modelling software (e.g., wireframe modeller, surface modeller)
Level 3	CAD/CAM systems software (e.g., user interface, data management)
Level 2	Graphics software
Level 1	Computer systems software (e.g., operating system, compilers)

controls the program as it works on the data and commands the peripheral devices. All computers from the very biggest "super-computer" to the smallest microcomputer have a CPU, and it is the difference in the type, quality, and quantity of the electronic circuits that go into the CPU that lead to different computers having different processing power. A wide range of CPUs is used in CAD/CAM. They range from "mainframes," "superminicomputers," "minicomputers" and "engineering workstations" down to "personal computers" and microcomputers. There is no ideal CPU for CAD/CAM, a given CPU may be suitable for one company of a certain size carrying out a certain application, and completely unsuitable for a company of a different size with a different application.

A company that does not want to purchase its own CPU for CAD/CAM work may use a timesharing service. This allows users to lease computer time and access the appropriate software in remote computers via communications lines. Another alternative for companies not wanting to purchase a CPU is to work with a CAD/CAM service bureau. These bureaus have their own CAD/CAM systems and carry out work for clients, or make their CAD/CAM installations available for use by clients.

2.1.2 Data Storage

Data and programs are stored by a computer in memory. The smallest unit of memory is the bit which can take a value of 0 or 1. This unit of memory is too small for practical use, so most actions in the computer take place on the byte (8 bits long) or the word. The length of a word depends on the type of computer, but is typically 8, 16, 32, or 60 bits. The size of memory is generally measured in units of KB (kilobyte, where 1 KB contains 1024 bytes) or MB (megabyte, where 1 MB contains 1048576 bytes). There is a wide range of types of memory. In general the shorter the access time (the time taken to retrieve information) of a type of memory, the more expensive it will be.

Main Memory

The main memory contains the program and the data currently in use. As the program should work as quickly as possible, the main memory should have a very fast access time. It is generally RAM (random-access memory) and is relatively expensive. For cost reasons, the main memory is kept as small as practical, and in a typical computer installation, it may represent less than 1% of the total memory available.

Mass Memory

Mass (or peripheral) memory is used to store the major part of the data and programs in a CAD/CAM system. Examples of mass memory are magnetic tapes, disks, floppy disks, cassettes, and paper tape. Magnetic tapes are useful for archiving data and for transferring information from one computer or machine to another. They have the advantage of being able to store large quantities of data at a low cost. However, the access time to a particular item of data can be very long as magnetic tapes are read sequentially. In contrast, magnetic disks are read by random access and have an access time measured in milliseconds. They are suitable devices for CAD/CAM where the need for fast interaction is so important. Disks come in various sizes, generally ranging from a few tens of megabytes up to several hundreds of megabytes.

Floppy disks are also available. These are small inexpensive magnetic disks with limited storage space. Cassette magnetic tapes also have a low storage capacity but are a convenient medium for temporary storage. Paper tape is also used to store information, in particular for NC machines.

2.1.3 Data Input

Information can be fed to the CAD/CAM system by a variety of input devices which enable the operator to select any of the available functions, enter texts and numeric data, change the display on the screen, input digitized data, or select a particular entity on the screen for treatment. Such selection by a pen, stylus, or cursor is termed "picking." A given input device is generally unsuited for at least one of the above functions, thus for ease of use, most systems have more than one such device.

Alphanumeric Keyboard

The most common device is the alphanumeric keyboard, similar to that used in a typewriter and through which the operator can enter commands and data. The keyboard is often used to enter precise data such as dimensions and other measurements. Some alphanumeric keyboards also incorporate a set of extra keys that can be used to carry out specific functions.

Function Keyboard

The function keyboard consists of a box equipped with a set of pushbuttons, of which there are generally 16, 24, or 32. The buttons can be labelled by overlays, and the overlays changed from one application to another. Sometimes, buttons can be illuminated by computer control, so that the operator can see which buttons are active. When a button is depressed, a preprogrammed function is immediately brought into use. The function keyboard can therefore give a faster access time to a function than the alphanumeric keyboard, on which several letters or keys would have to be chosen to select the same function.

Menus

The alphanumeric keyboard and the function keyboard offer two ways to enter commands and select system functions. An alternative technique is the use of "menus." A menu of functions can be displayed on the screen, and the required function selected or "picked" by the user. A menu can also be presented on a plastic or paper sheet. The sheet is divided up into a set of zones. The user picks the zone corresponding to the required function.

Graphics Tablet

The graphics tablet (or data tablet or tablet) consists of a rectangular grid of wires underneath a plane, electronically sensitive surface. Graphics tablets range upwards in size from about 20 X 20 cm to the large digitizers used for digitizing full-scale drawings. The location of a picking device positioned on the surface of the tablet is determined from the pattern of signals on the wires. The tablet is used in two ways. First, it can be used as a digitizer. In this case a drawing or menu can be placed on the surface. The picking device is then used to pick up coordinates from the drawing or to select a zone from the menu. Alternatively the tablet can be used to select a line or character on the graphics screen. This technique involves mapping the tablet surface to correspond to the display screen surface, with a cursor (for example crosshairs) being displayed on the screen to represent the corresponding position of the picking device on the tablet surface. As the picking device is moved round the tablet, the cursor on the screen follows its movements. An object on the screen can be selected by moving the picking device until the cursor coincides with the object position on the screen, and then signalling with a switch that the cursor is in the required position.

Picking Devices

The two picking devices most often used with the graphics tablet are the puck and the stylus. Crosshairs at the center of the puck are used to define a position, and a set of pushbuttons on the puck are used to signal information, for instance that the crosshairs are at the required position. The stylus is similar in shape and size to a ballpoint pen. Its "nib" is used to select a position, and a switch at the other end is used to signal position acceptance.

In the absence of a data tablet, other devices are used to indirectly control the cursor on the screen, thus bringing it to a position where it is used to select data or commands. These devices include thumb wheels (with which the operator can generate separate X and Y cursor movements) and the trackerball and joystick which can be used to generate simultaneous X and Y motions. A mouse can also be used as a cursor control device. The light pen is a picking device that works directly on the face of the graphics screen. The touchscreen is a picking device that may be used increasingly in the future. A recently developed but currently little used data input device is the voice recognition system. This allows commands to the CAD/CAM system to be entered verbally through a microphone.

2.1.4 Data Output

Graphics Terminal

The graphics terminal should provide several functions including rapid generation of a steady and clear image on the screen, the carrying out of some processing itself (i.e., local processing) and the handling of the communication of data and commands between the user at the terminal and the main CAD/CAM software in the main computer. The more a terminal can do itself independently of the computer, the more it is regarded as being "intelligent." In contrast, a "dumb" terminal is one which does no more than display an image on the screen as a set of lines, with all other functions being carried out by the main computer. Between the dumb terminal and the intelligent terminal is the "smart" terminal with its own microprocessors capable of carrying out special graphics functions such as filling defined areas on the screen with a given color, or drawing a circle with a known center and radius.

The basic element of the most common graphics screens is the cathode ray tube (CRT). Graphics screens can be classed in different ways. First, the choice of the phosphor for the screen defines the need to "refresh" the image so that it will appear to be steady and without flicker. The light from some phosphors dies out within milliseconds of the beam moving to another location. With other phosphors, light will continue to be emitted for several seconds, and some allow the image to be held indefinitely until it is intentionally erased. The process of rewriting the image so that it appears steady to a user is called refreshing the image.

There are three major classes of screen—vector refresh, raster refresh, and storage. Two of these (vector refresh and raster refresh) use phosphors that have to be refreshed many times per second while the storage screen uses longer-life phosphor elements that are maintained illuminated by a secondary beam.

Graphics screens can also be classed as those using stroke-writing techniques and those using raster scan techniques. In stroke writing, the beam is moved from one X, Y position on the screen to another. During one movement, the beam will be either on or off. To draw a large Z on the screen, the beam would be turned off and moved from wherever it was to the top left of the screen. Then the beam would be turned on, moved to the top right, the bottom left and then to the bottom right. To draw a different picture, the pattern of beam movement would be different. However, in the raster scan technique, the beam is always moved in the same pattern regardless of the picture to be displayed. The beam starts at the top left of the screen and carries out a sweep of the first horizontal line. It then moves back to the left of the screen and carries out a sweep of the next horizontal line. This process continues until after reaching the bottom line and carrying out a horizontal sweep, the beam returns to the top left of the screen and the whole sequence is repeated. Stroke writing generally offers a higher resolution image and higher addressability for X, Y positions than the raster technique. Screens are often described by their number of addressable positions. On a 1280 × 1024 screen, 1280 positions can be addressed on each of 1024 horizontal lines.

In the 1960s the first CAD/CAM graphics terminals used predominantly stroke writing and refresh technologies. Vector refresh screens as they are known, are still in use. They give a bright, clear, and crisp image. Since the image is constantly being refreshed it may be modified very easily, and can even be used for giving an impression of animation. One problem that arises with these screens is that display of a lot of information (i.e., a lot of vectors), may cause the image to flicker as the phosphor dies out before it can be refreshed. Another problem is the relative difficulty of generating colors (rather than just green and black, or white and black), although their major disadvantage for most applications in a medium-sized engineering company may well be their relatively high cost.

Light pens work best on vector refresh screens. They can be used both to select existing data on a screen, and to enter new data. To select an object, the light pen is pointed at it, and detects the light of the beam redrawing it. Before a light pen can be used to create a new object, a cursor is made to appear on the screen. The light pen is then used to move the cursor round the screen and draw the object.

Storage tube screens were introduced in the late 1960s. They use stroke-writing techniques and long-life phosphors. The tubes have two main electron guns, one for writing and one for flooding. After the image has been written, the phosphor continues to luminesce, thus there is no flicker. The displayed image can be added to by further writing, but there is no selective erase feature. Instead, to delete a part of the image, it is necessary to flood the picture (causing a bright flash), thus effectively cleaning it, and then redraw the modified picture. Other disadvantages of storage tubes are their low brightness and contrast. However, due to their low price they have often been used in CAD/CAM applications. Some storage tubes offer a limited refresh capability to enable menus to be displayed on the screen and to enable a little animation.

The third major class of graphics screen, the raster refresh screen (often just called the raster screen) began to make a sizable impact in engineering companies in the early 1980s. Each horizontal line of a raster sweep is made up of a certain number of points. The contents of each point (or pixel) on the screen must be defined and stored in a memory called the bit map. As the beam scans the entire screen from top to bottom, it is turned on or off at each pixel as a function of the on-off pattern stored in the bit map. For a black and white screen, each pixel can be black (beam off) or white (beam on). This information for the pixel can be stored in one bit (e.g., 0 representing black and 1 representing white). The bit map for a screen with 1000 points per horizontal line, and 1000 horizontal lines therefore requires 1,000,000 bits of memory. A bit map for a black and white screen is called a memory plane. Additional memory planes can be used to provide information on color at each pixel. Two memory planes can define 4 colors, 4 memory planes, 16 colors and so on.

The cost of memory tends to dominate the price of raster screens. To keep prices low, memory may be kept to a minimum but this means only a limited number of points and lines being available which in turn results in comparatively low resolution of the image. This is particularly noticeable on diagonal lines which have a jagged or staircase appearance. Various techniques, such as antialiasing, are available to reduce line jaggedness.

Raster screens can display large amounts of information without flicker, have selective erase features and offer low cost color capabilities. With a high enough resolution, they are currently the most suitable displays for use in CAD/CAM engineering applications.

Alphanumeric Screen

The purely alphanumeric screen has a limited role to play in CAD/CAM. It is generally only used in association with a graphics screen. It is used for echoing data and commands input by the user, and for displaying messages and the results of calculations.

Plotter

An image on a graphics screen is only of use while a user is in front of the screen. For further use at the desk or as a permanent record, a hard copy is needed. Sometimes a "local" hard copy unit is provided with the screen. This gives a quick but fairly low quality picture of the screen. A more accurate but slower technique is the use of a plotter. The two basic types of plotters are electrostatic plotters and pen plotters. Pen plotters use stroke-writing techniques, whereas electrostatic plotters use a raster scan.

Drum Plotter

The most widely used pen plotter is the drum plotter, in which paper is wrapped around a drum which is rotated. The pen (of which there may be several) is mounted on a gantry above and across the drum. The pen can therefore move along the length of the drum. It can also be moved up or down (off or onto the paper). When the pen is lowered onto the surface of the paper, a line is drawn on the paper in a direction given by the combined movements of the rotating drum and the pen. The length of the drum gives the width of the drawing, which can be of any length. The drum plotter is accurate, reliable, and takes up little space, but it is relatively slow compared to an electrostatic plotter.

Flatbed Plotter

Flatbed plotters range from very large sophisticated drafting machines down to small, cheap devices. In all cases though, the paper (or other drawing medium) is fixed to a flat drawing surface, while the pen can move along a gantry that is driven in a direction perpendicular to the movement of the pen. Some flatbed plotters offer a choice of pen nibs and ink colors. The more sophisticated flatbed plotters are used where extremely accurate drawings are required.

Electrostatic Plotter

Electrostatic plotters can produce drawings many times faster than electromechanical pen plotters, but they tend to be more expensive and the quality may not be so high. A hidden cost of the electrostatic plotter can be the use of the computer to translate "a drawing" into a form in which it can be transmitted to the plotting head. As in the drum plotter, the paper's motion provides the lengthwise axis of motion. The actual plot is performed in two phases. First, as a function of the data to be plotted, selected nibs embedded in a stationary writing head running across the width of the paper create minute electrostatic dots on the special charge-sensitive paper. The paper is then exposed to a liquid toner thus producing the corresponding line of the plot. An advantage of electrostatic plotters is that they can also be used, if so required, as line printers. Some electrostatic plotters can produce a drawing in color.

COM Plotter

COM (computer output to microfilm) plotters are available for putting output graphics directly onto microfilm for convenient storage.

Other Plotters

Laser plotters, inkjet plotters, and thermal transfer plotters are currently used very little in mechanical CAD/CAM.

2.1.5 Communications Equipment

In most CAD/CAM systems, the peripherals (data storage, input and output devices) are situated near to the computer and are connected directly to it. Communications equipment becomes necessary when peripherals are situated far from the computer.

The earliest form of computer communications of concern to CAD/CAM users was that of telephone links to a remote computer. Such links are accomplished using modems. The baud rate (or transmission rate) is given by the number of bits of information communicated per second. In most cases, alphanumeric information can be transmitted at speeds as low as 300 to 1200 baud without causing user frustration. However, for graphics work, speeds over 9600 baud are generally needed. Special arrangements can be made with the telephone authorities to transmit data at speeds up to several million baud.

Local area networks (LANs) which transmit data at up to 10 million baud are often used to connect processors and processor-controlled equipment in geographically limited areas. (Ethernet is an example of a LAN.) Dedicated networks have been developed for specifically communicating between processors produced by the same computer manufacturer. Other networks exist to communicate between computers from different manufacturers spread over wide geographical areas.

One of the major problems encountered with networking is that of making equipment (e.g., PLCs, computers, workstations) from different manufacturers communicate with and understand each other. International organizations are working to establish standard protocols.

2.2 SOFTWARE

The six levels of software within a CAD/CAM system will now be described in detail. They are illustrated in Table 2.2.

2.2.1 First-Level Software

The basic arithmetic operations of a computer are addition and subtraction. Computer programming languages have been developed to allow programs to be written to control the execution of the computer using higher level constructs. One such program, called the operating system, and supplied with the computer, controls the CPU, and commands and communicates with the peripherals. Other programs, called compilers, are also supplied with the computer. The FORTRAN compiler, for example, takes a program written in FORTRAN source code and translates it into machine code statements that can be executed by the CPU.

Programming languages such as assemblers only offer programmers very low-level constructs. Others such as FORTRAN and COBOL offer higher level constructs. Up to 1985, most CAD/CAM systems were written in FORTRAN.

The operating system and the compiler can be regarded as the basic or computer system level of software needed by the CAD/CAM system. They are not really part of the CAD/CAM software since exactly the same operating system will be supplied by the computer manufacturer to a user who buys the same computer for purposes other than CAD/CAM. However, the CAD/CAM system could not work without the operating system and the compiler.

2.2.2 Graphics Software Level

The second level of software within the CAD/CAM system is the graphics software that controls the graphics terminal.

The graphics software has several tasks, to name a few, it communicates with the main computer, it manages the input/output devices such as the keyboard and the puck that are associated with the screen, it "draws" the image on the screen. Graphics software, like programming languages, comes with different degrees of sophistication. The least sophisticated understand little more than commands to move the beam from one point to another, or to switch the beam on and off. The programming instructions in such languages are very specific to the particular screen, thus a program written to run on one screen would not be usable on another screen.

In the more sophisticated graphics languages, an attempt is made to ease the difficulty of preparing programs, by offering the user a library of routines capable of carrying out higher level functions such as drawing a line with a given width and style, using colors, generating polygons, erasing and then recalling lines, reading a value input at the keyboard, and so on. These routines can be called by the programmer from a program written in a high-level programming language such as FORTRAN. Unfortunately, the majority of these sophisticated graphics languages are screen dependent, thus a program written for one screen will not run on another.

In an attempt to free the programmer (and user) from being tied to a single screen, attempts have been made to develop an international standard defining efficient device-independent graphics software containing all of the necessary procedures. Two standards have been proposed (CORE and GKS) although neither is currently suitable for all applications.

Very sophisticated graphics software can also be used to improve the understandability of an image, for instance, the removal of lines representing parts that would not be visible to an observer in a given position, or color shading of an object under various lighting conditions.

2.2.3 Third-Level Software

The third level of software is made up of the general CAD/CAM software that will be used by all the individual CAD/CAM applications. This level includes the user interface, data management, and data exchange software.

The user interface allows the user of a CAD/CAM system to communicate with any of the various CAD/CAM application programs. It contains functions such as the control of the means by which the commands are entered (e.g., by screen menu, by tablet menu, by keyboard), the input of texts and numeric values (by keyboard, by function keys, or from the tablet) and the output of texts and numeric values (e.g., at the screen, to a file, on a listing). It is also the part of the system that is used in prompting the user, displaying help messages, selecting a particular entity among all those available, digitizing a point. It will be seen that these functions are common to all CAD/CAM applications. They are as applicable to preparing a tool path or a finite-element mesh as they are to drafting. Similarly, since the various CAD/CAM applications within a CAD/CAM system access the same data, the data management function is common to the system, rather than to the specific application. Data exchange software, used to exchange data between CAD/CAM systems is also at this level.

2.2.4 Product-Modelling Software

The fourth level, containing the geometric modelling software, is one of the most important. The quality of the geometric modeller will define just how powerful the individual CAD/CAM applications can be, and to what level integration between the applications can take place.

The geometric modeller is the part of the CAD/CAM system that is used to define or represent a part as a computer-usable model. One way to classify geometric modellers is between those that build a two-dimensional (2D) model of a part and those that build a three-dimensional (3D) model. Two-dimensional geometrical modelling is basically computerization of the drafting process. For each view, the model in the computer knows the position of the various lines and arcs that would normally appear on a drawing, but there is little connection in the model between corresponding features in different views. Although such a modeller is eminently suitable for two-dimensional applications such as electric circuit design, it is not always as suitable in mechanical and manufacturing engineering where the object being modelled is often truly three-dimensional. However, 2D geometric modellers are used in mechanical and manufacturing engineering. They are really computer-aided drafting systems, and as such can give rise to increased speed of production of drawings.

The three most common types of 3D geometric modelling are known as wireframe modelling, surface modelling, and solid modelling. In the wireframe approach, the user creates a 3D model using points, lines, and arcs to describe the edges of an object. This is a very fast method of creating a model and can give a quick response time even on a microcomputer. However, it suffers from the major disadvantage that what lies between the edges (or wires) is not defined, thus some of the information needed in some applications may be lacking.

Surface modelling aims to provide a complete definition of the surface of a part. Thus, whereas the wireframe modeller would represent a cube as 12 lines, the surface modeller would represent it as 6 surfaces. One of the great advantages of surface modelling is in modelling sculptured (or doubly curved) surface parts such as car bodies, helicopter rotor blades, aircraft wings and turbine blades. Complete geometrical information is available for each point on the surface and can be used in applications such as NC programming and mesh generation. For simple shapes, a surface model may contain unnecessary information, this "overkill" is paid for by the user in a longer data entry time and a slower response time.

Solid modelling aims to give a complete description of a part including not only its edges and surfaces but also what is inside. There are several classes of solid modeller. One type, known as constructive solid geometry (CSG) involves the combination of primitive solid geometric shapes (e.g., cube, sphere) by Boolean operations. As an example, assume that a cube is a solid primitive known to the modeller. If a small solid cube is subtracted from the middle of a larger solid cube, the result will be fully understood by the modeller, as it will know that there is a hole in the middle of the larger cube. Another type of solid modeller known as a boundary representation (B-rep) modeller constructs a model by piecing together the external surfaces of a part.

Since the solid modeller is supposed to contain all the geometric information on a part it should be the most powerful of the 3D modelling techniques. However, at the present time, such modellers are used very little in mechanical and manufacturing engineering. Among the reasons for this can be cited the slow and cumbersome input to some of these modellers, their inability to model sculptured surface parts and their need for a powerful computer to ensure a fast response time.

Some of the major differences between CAD/CAM systems derive from the differences in their geometric modelling capabilities. Not only do wireframe, surface, and solid modellers work with different descriptions of the same part, but often modellers of the same type (e.g., wireframe) available in different commercial CAD/CAM systems, create and store their models differently. A problem arises when an attempt is made to transfer data from one modeller to another either within the same company, or from one company to another. To overcome this problem various standards (such as IGES, VDA, SET) have been proposed for the format in which data should be transmitted between CAD/CAM systems. The types of data to be transferred from one system to another include geometric data (e.g., lines and circles), nongeometric data (e.g., dimensions and texts), drawing and display data (e.g., line patterns and thicknesses), and relationships between these (e.g., connectivity and assembly structure). Taking the IGES standard as an example, data from CAD/CAM system A would be converted into the IGES format (by an IGES preprocessor) and transferred in this form to CAD/CAM system B. This system would have a converter (an IGES preprocessor) to convert data from the IGES format to the format of its own geometric modeller. At present, none of the proposed standards cover all applications.

2.2.5 CAD/CAM Applications Software

At the fifth level of CAD/CAM software is the CAD/CAM applications software. It will make use of all the underlying levels. Before describing the applications software in detail, the original definition of CAD/CAM should be recalled. The three major concepts involve interactive graphics, geometry modelling, and reusable data. Starting at the beginning of the process, the design, the geometry will be modelled and stored in a data base. A CAD/CAM application is basically any design engineering or manufacturing engineering process which can benefit from the use of interactive graphics techniques and the use of part geometry as contained in the data base. Table 2.3 shows some of these processes.

Once the geometric model of a part has been completed it is possible to calculate some of its geometrical characteristics. Just how many characteristics can be calculated will depend on the richness of information made available by the modeller. With surface modellers and solid modellers it is generally possible to calculate volumes, surface areas, moments of inertia, lengths, and angles.

One of the uses of CAD/CAM at the analysis stage is in preparation of the mesh for finite-element analysis of a part. Manually this is a time-consuming, tedious, and error-prone task. With CAD/CAM, the mesh can be generated directly from the geometric model of the part. For example, if a surface modeller has been used the coordinates of any point on the surface will be known. The user uses a mesh generator to define the node points for the elements and their connectivities. The mesh, loads, and other parameters are then input to an analysis program (not part of the CAD/CAM system). The data output from the program can then be displayed on the graphics screen superimposed on the original data.

It is also possible to use CAD/CAM in kinematic analysis. Starting from the geometry of the part and adding information on the position of hinges and other linkages, the motion of a part can be analyzed, for example, to ensure that it does not interfere with other parts.

Table 2.3   The Impact of Geometry Modelling on Some of the Design Engineering and Manufacturing Engineering Functions That Will be Affected by CAD/CAM

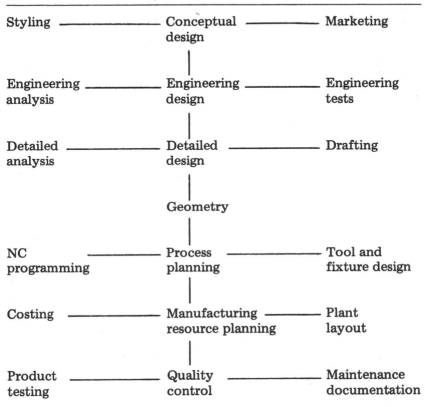

Styling	Conceptual design	Marketing
Engineering analysis	Engineering design	Engineering tests
Detailed analysis	Detailed design	Drafting
	Geometry	
NC programming	Process planning	Tool and fixture design
Costing	Manufacturing resource planning	Plant layout
Product testing	Quality control	Maintenance documentation

Drafting is the most widely used CAD/CAM application. A wide range of drafting software is available. Some systems do little more than replace the drawing board surface by the screen surface, leaving all but line and character drawing to be carried out by the user. At the other end of the range are systems with functions such as automatic cross-hatching and dimensioning.

In the manufacturing engineering area, CAD/CAM applications include generation of bills of materials, geometry modelling and drafting of tools, and generation of process plans. Another application is in programming NC machine tools. The program can be developed using part geometry available in the data base. For some part geometries, tool paths can be displayed and checked on the graphics screen thus reducing the time actually spent on the machine tool for verification.

Numerically controlled programming is generally taken to mean the programming of numerically controlled machine tools. The programming of robots is considered a separate problem and is often carried out in a teach mode, during which the user physically moves the robot through the steps of an operation. An alternative is to define the robot path in a CAD/CAM robot programming application package. This would use part geometry coming directly from the data base.

Within mechanical and manufacturing engineering there are many other CAD/CAM applications. For example, wiring diagrams can be built up using CAD/CAM techniques and NC drill and insertion tapes prepared from information in the data base. Similarly piping can be designed with CAD/CAM techniques and then manufactured with NC pipe bending machines. CAD/CAM can be used in planning plant layouts and in simulating automatic transfer machinery behavior.

2.2.6 User-Developed Software

The CAD/CAM applications software described above is made available by the CAD/CAM system vendor. However, there are generally some industry-specific or company-specific application functions not supplied by the vendor. These will have to be written (or at least specified) by the user, and they form the sixth level of CAD/CAM software. Generally, the vendor makes available access points in the CAD/CAM system at which user-developed software can be added.

2.3 PUTTING THE INGREDIENTS TOGETHER

It will have been seen that there is a wide range of hardware and software available for use in CAD/CAM. The number of possible configurations is thus very high. However it is possible to divide the systems available on the market into a limited number of categories although the edges of the categories are often hazy and overlapping.

In the "service bureau" category, users with terminals on their own sites can run CAD/CAM packages on remote computers belonging to service bureaus.

In the "centralized mainframe category" a company runs a large number of terminals (say 20 or 30) off a mainframe computer.

In the "distributed computers" category a company runs clusters of small numbers (say 3 to 6) of terminals off each of a set of superminicomputers. Advantages of this solution are that each computer can be sited near its users, computers can easily be added to expand the system, and that even if one computer is not working, the other users are not affected. On the other hand, without good networking and distributed data base software this solution may not be totally effective.

In the "turnkey" category, a complete system comprising all the necessary hardware and software is supplied by one vendor. In principle it can be operated by the user immediately after installation "at the turn of a key." It should not therefore require a high level of computing expertise from the user (although it will require a high level of CAD/CAM expertise). As all hardware (computer, terminals, plotters, etc.) is supplied by one vendor, the supplier can optimize use of various devices and software in the system.

In the "software package" category, a vendor sells only the CAD/CAM software and not the hardware. In this case the vendor will generally supply the software to run on several types of computer and graphics screen. Although a user may appreciate the potential of running the same software on various computer systems, the software vendor has less freedom to optimize the software to be more efficient on particular devices.

In the "engineering workstation" category, only one graphics terminal runs off each CPU. Engineering workstations can be bought turnkey (i.e., with all necessary hardware and software). Alternatively, users may buy individual engineering workstations and then purchase the other CAD/CAM hardware and software independently.

In the mid-1980s, personal computer-based systems are inexpensive and easy to use. They are mainly limited to drafting and training activities because they lack the processing power and data management capabilities to handle the complete range of CAD/CAM applications. However, during the period 1985–1990, as personal computers become more powerful, they will be able to offer the same applications as engineering workstations.

2.4 TRENDS

CAD/CAM appears to have gone through three distinct phases. In the first phase (the 1960s) systems were developed by high technology companies (e.g., aerospace and automobile companies) for their own use. They were generally based on mainframe computers and vector refresh screens.

In the 1970s, turnkey system vendors started to develop and supply CAD/CAM systems to user companies. These systems were based on minicomputers and used storage screens.

In the 1980s, the trends so far have been to the superminicomputer and the raster screen on one hand, and on the other hand to the engineering workstation (i.e., a powerful microcomputer combined with a raster screen). Although a few user companies continue to develop their own systems (in particular their own software), the vast majority of user companies use CAD/CAM software supplied by turnkey or software package vendors.

3

SELECTION OF A CAD/CAM SYSTEM

Before the procedure for selecting the right CAD/CAM system for a company's applications is described in detail, a description is given of how a fictitious company made its choice.

3.1 FICTITIOUS SYSTEM SELECTION IN PITTSBURGH

Golden Triangle International Furniture Inc. of Pittsburgh, Pennsylvania (and its equally fictitious staff) found itself losing orders to overseas competitors. Its President, Fritz Froehlicher, had emigrated from Switzerland in the early 1950s and had founded the company in 1954. A few years ago, he had handed over the everyday running of the company to Wayne Grotzer. Wayne had been born and bred in Pittsburgh, and had spent most of his life in the steel industry. His marriage to Fritz's daughter, Heidi, was responsible for his move to the furniture industry.

One day, while Fritz was 'talking to Wayne about how the company might try to fight off foreign competition, he suggested that they might try automating the plant. "Great," said Wayne, "then we'll get back to being Number One." Four years ago, Wayne had been to business school for a refresher course. Since then he had implemented all the new management methods he had learned there. He had also diligently read that book that had topped the "Virile Management Booklist" for 29 consecutive weeks, but he still could not understand why orders were being lost. He did not remember reading about automation, and at business school there had been no mention of such an unsavory topic. Perhaps his father-in-law was right. Maybe there were some problems on the shop floor that could be put right with a few computers.

On Monday morning, Wayne met with his management team. That Monday all five members were present. David Olafson, the Finance Director, was responsible for everything apart from furniture design and production. That made him an important part of the management team. Arthur McIlroy was the Design Office Manager. He had been recruited from England many years back when the U.S. aircraft industry was booming. Although a stress engineer by training, at heart he was a craftsman. He had developed excellent all-round management skills while in the aircraft industry, and when the previous manager of the Design Office had retired, Fritz Froehlicher had recognized Arthur as the ideal replacement. Werner Schumann ran the Tooling and Cutting Shop. After attending the best engineering schools in Germany, he had carried out postgraduate studies in the United States and decided to settle in Pittsburgh. Had he returned to Germany, by now he would probably have had a top managerial position at one of the automobile companies. Claude Le Paysan ran the Assembly Shop. While he lacked the academic background of the other team members, Claude had worked his way up through the ranks.

The sixth person present at the meeting was Anna Bellinaso, who ran the Finishing Shop. Another convert from the steel industry, she had been with Golden Triangle for nearly five years. Anna had been worried recently as there hadn't been much work passing through the Finishing Shop.

Wayne started the meeting brutally: "Look you guys, either we automate, or we liquidate, or we emigrate. I want your suggestions by next Monday morning." At first, they were silent, Claude Le Paysan spoke first "Automation is a great idea. You'll have my suggestions tomorrow morning," David Olafson spoke next, "We've got some spare capacity on our computer system down in Finance. I'll talk to some of our programmers and see what we can do." Werner Schumann and Arthur McIlroy were preoccupied with some tools that had to be changed for a modified design and Wayne's suggestion caught them by surprise. They knew little about automation. Anna Bellinaso was surprised and distressed. She envisioned the worst aspects of automation, including robots replacing her staff. "Don't robots cost a lot?" she asked.

Wayne Grotzer did not know how much robots cost and thought it best to change the subject. Later that day, while lunching with his father-in-law he said, "Fritz, I still haven't solved this question of automation. I don't think we need robots in the Finishing Shop, but we could put them into the Assembly Shop during the summer vacation shutdown." While Fritz Froehlicher had every confidence in Wayne, he felt he needed some guidance now. "Look Wayne," he said, "before you get down to the detailed planning, let's set out our objectives and try to develop a suitable strategy. First of all, see Dick Nelson in Marketing and find out why we're losing market share."

Dick Nelson was surprised to be called to Grotzer's office that afternoon but quite capable of telling him why market share was falling. "One reason is that we take too long to make proposals. It takes us nearly three weeks, but some of our competitors reply in three days. Another reason is that it takes us too long to build furniture, and often parts don't fit very well—that gives us a bad image." Wayne Grotzer was relieved to hear that he wasn't going to need to put robots in the Finishing Shop. Before Dick Nelson left the office, Wayne asked him how he saw the company evolving over the next 5 years. Dick replied, "I've often thought about that one. Our competitors are producing higher quality furniture than ever before. They are making more models than ever and they are changing their models more frequently. If this company doesn't change its way of doing business, it will be out of business in 5 years."

That evening at dinner Wayne was able to tell Fritz that the company's objectives had to be to reduce time cycles in design, cutting, assembly, and finishing, and to increase quality. Fritz was pleased to hear this, since he had done a little private research and come to much the same conclusion. However, he had also tried to find out what their competitors were doing and it seemed that although they were going for quality, quick delivery, and frequent model changes, they appeared to be restricting their range of basic models. Fritz saw the possibility for Golden Triangle to gain a competitive advantage so he added one more objective, the need to produce a wide range of models. Later that evening, it was agreed that Fritz would find out what resources the company could invest in automation, while Wayne would continue to try to discover which automation technologies would help them meet their objectives.

The next morning, Wayne Grotzer found a 40-page report from Claude Le Paysan on his desk. As he skimmed through it, two points stood out—first, the company needed a Director of Automation, and second, the company should buy assembly robots linked to a CAD system. This would give a productivity increase of 10 to 1, but 90% of the people in the Assembly Shop would be laid off. As Wayne Grotzer was wondering what should be done, the phone rang. It was David Olafson, the Finance Director. His systems team had done some analysis of the automation problem and thought that it could be handled with the existing computer. All they needed now was a new terminal and three more programmers—did Wayne agree? "Don't take any action yet, David," said Wayne "I'll call you back later." As he wondered why any mention of computers or automation seemed to cause instant madness and disorder, he saw that Claude Le Paysan had included the name and telephone number of the CAD system salesman in his report.

Wayne was astonished to find that the salesman was none other than his old schoolfriend Bill Laritz. He picked up the phone and called him. "Hi Bill, this is Wayne Grotzer. I didn't know you sold computers, I thought you worked for a domestic appliance company." "I did, Wayne, but these CAD companies offer great opportunities, so I moved over." "Well Bill," said Wayne Grotzer, "I run Golden Triangle International Furniture Company now, and you're the guy to tell me how to automate my company." After a long silence, Bill Laritz replied, "You were always a good and trustful friend, Wayne, so I'll not give you the usual pitch. I've only been here four months so I can't tell you how to automate your company—I don't even know how to use this system." "But, Bill, you told my Assembly Shop Manager that this system gave a productivity increase of about 10 to 1." "Sure," said Bill, "only last week a guy came in here, sat down in front of the CAD screen and designed a circuit in 5 minutes. Without the system he reckoned it would have taken him about an hour—so that's a productivity ratio of 12 to 1." "Look Bill, don't think I'm being aggressive, but I make furniture not circuits—and who was that guy? My designers can't even play video games, they won't be able just to sit in front of a screen and design something." "I'm sorry, Wayne, but as I told you, I'm new here, so I can't answer all your questions. However, I'm reading a good book on CAD at the moment. When I've finished it, I can send it over to you. There's a good CAD course on this week at the university, you could go to that as well."

Wayne went to the course, and bought himself a copy of the book. He was ready for surprises at the next Monday morning meeting of the management team, but even so, found himself astounded. Claude Le Paysan had brought along 7 copies of his report. David Olafson had brought along some forms that he wanted Wayne to sign so that the new project could get under way immediately. Werner Schumann was proposing a numerically controlled saw that could cut wood, metal, or cardboard and had an 18-month payback. David Olafson said that any payback period longer than 12 months was unacceptable (although he did not say what the payback period for his project was). Arthur McIlroy was proposing a stress analysis computer program that would allow furniture designs to be analyzed by computer, thus obviating the need to build a prototype. Anna Bellinaso had prepared several pages of tables showing the cost of a painting robot. However, by changing the amortization procedure, the number of shifts worked and the interest rate, the cost could be varied between $5.25 and $120.76 per hour.

"Thank you for all your contributions," said Wayne "we've made a lot of progress. From my side I should tell you that I've been refining my ideas on our objectives. I'm also in the process of evaluating the resources that we could commit to automation. However, I believe that we're still missing the global picture of how our business works today, and within that overall picture we need to identify the special points where automation can do most to help us attain our goals. I'm setting up a task force to get that global picture and to find those special points," he continued, "I've decided to lead the task force myself. I'll be co-opting one person from each of your shops. Since none of us know much about automation, I'm also getting an experienced management consultant in to tell us which automation technologies we could profitably use, their advantages and disadvantages and how we should implement them. Rick Nelson from Marketing should be with us and we'll also need someone with a good financial background—David, do you see anyone suitable?" David Olafson saw no reason to put forward his own name, so he suggested Ed Falcon from Accounting. "That's a good choice, David" said Wayne Grotzer, "we need someone who is not going to get overwhelmed by words like productivity, payback, amortization, capital investment, internal rate of return, and so on. Those are just words that are used to express relationships about parameters of the business. They are often used as crutches by people who don't understand engineering and manufacturing. If we don't understand the way our engineering and manufacturing process works, and the automation technologies available, crutches like that won't help us to find the best strategy."

The task force met many times, analyzed the way the company worked, visited other companies, went to conferences, visited automation technology vendors, and so on. After a few months it became clear that among the most interesting automation technologies for the company were CAD, NC cutting tools, a production planning and control system, and quality control equipment. The task force then produced a report in which it was recommended that in view of the company objectives as originally formulated, and as a function of the financial resources available, the two automation technologies with the most potential for the company were CAD and NC cutting tools.

As Fritz Froehlicher and Wayne Grotzer discussed the report, they felt proud that in less than four months their company had advanced so far along the road to finding the best automation technology to be implemented. Now, they said, it should be simple to select the best systems in the chosen areas. As the company objectives and available resources had been clearly stated, Fritz Froehlicher decided that his involvement was no longer needed. As Wayne Grotzer was very busy, they decided that the task force leader for the next stage would be Rick Nelson from Marketing. He was respected throughout the company and having worked in several departments, knew it well. In addition, as he would not be directly affected by the outcome of the selection, he could not be accused of being partisan. Wayne would give him all the assistance and support necessary. To ensure that the most suitable solution was found for the entire company, and not just for a particular shop, Fritz and Wayne ensured that the task force was really representative of the whole company.

Along with Rick Nelson from Marketing and Ed Falcon from Finance, they would include the management consultant who had helped them so much in selecting the appropriate automation technology. It was decided that the task force would also include Malcolm Byers, a senior craftsman from the Assembly Shop and Paul Valdas who had been a NC programmer with his previous company. Arthur McIlroy and Anna Bellinaso would complete the task force.

The first meeting of the new task force was very difficult for Rick Nelson. Paul Valdas told the meeting how the NC system that he had used before had worked well. Malcolm Byers recommended Wonder Robot, a robot that had the advantage of being small enough to be installed in an office, but was still in the development stage. Arthur McIlroy had been to another demonstration of the stress analysis program that he favored, and was sure it would change the fortunes of the entire company. Ed Falcon had prepared a long presentation entitled "Finance and Automation." Anna Bellinaso was unable to stay to the end of the meeting because of production problems in the Finishing Shop. Rick Nelson just about managed to bring the meeting to a close and fix the following meeting for the next Friday.

Rick went back to his office, and it was there that Wayne Grotzer found him, staring blankly out of the window. "I'm sorry Wayne, but I don't think I can lead this task force." "Come on, Rick, everyone gets a touch of the CAB (computer-aided blues) now and again. Sit down and let's talk about this. You know, Rick, one thing I noticed when we ran our first task force was that to start with everyone wants us to use a technology that he can understand, and feels he can master. You have to get them away from that attitude. One of your first jobs as task force leader is to pass on to the team your knowledge of the company's objectives in setting up the task force. Get them to think globally and act together, not as a set of individuals. Tell them that it's normal that at the moment there are a lot of things they don't know, but that one of the reasons for setting up the task force is to increase the overall knowledge of the company—and in the long run the company is just made up of people like themselves. Remind them that it is important for the company to reduce time cycles in design, cutting, assembly, and finishing. They probably understand just as well as you and I do how important it is to increase quality and to produce a wide range of models. Let them know the kind of resources that we have available and why we are now looking in detail at the areas of CAD and NC cutting tools. You know, if you share the information that you have available with your people they are much more likely to work for you and with you than if you hold back information or give false information. Those people and the information in their heads are this company's major resource—we have to use them as efficiently as possible, not only for the sake of the company but also for their own sake. Get them to realize that they can't choose systems until they know how our operations work and how they could be impacted by CAD and NC. They need to know the basics of CAD and NC. Book them in to a good introductory course, then take them to see some users. If there are some good exhibitions on, send them along. It's very important that the task force understand what we do here and that they learn enough about the basics before the vendors start selling to them. If they don't have that level of understanding, there's no way we can make the right choices. Don't look so glum, Rick, you can do it."

Rick went away and started to plan how he was going to run the task force, and exactly what each of its members was going to do. The meeting that Friday was completely different from the previous one. After explaining the objectives again, Rick told the task force that in 6 months time they had to present a report to Wayne Grotzer. The report had to list in detail the requirements of the company in CAD and NC, and explain the task force's preferred solution. Before they could protest, Rick handed out the first list of tasks: Anna Bellinaso and Ed Falcon were to go to a course on Automation Finance and Productivity at a Midwest university. Paul Valdas and Arthur McIlroy would go to Detroit for a major exhibition, and to visit some jobshops using CAD and NC. Malcolm Byers, the senior craftsman, was to visit Pittsburgh area companies that were already using CAD and find out what they were doing with their systems. The management consultant and Rick Nelson would start to detail the way in which information and work was created and exchanged within the different areas of the company. Rick also handed them two books each and said that by the next meeting, in three weeks' time, everyone should have read the books. At that meeting, each of them was to make a 30-minute presentation on what had been learned so far.

The following three weeks were difficult and several members of the task force received a lot of help from Rick when they went to see him about various problems that arose. Surprisingly enough, it was Rick himself who ran into the biggest problem. In the Assembly Shop, Claude Le Paysan was too busy to provide the information Rick needed. Eventually Rick asked for Wayne Grotzer's help, and together with Le Paysan, they managed to get a fairly accurate picture of the process.

As time went on, and the task force members continued their work, it was clear that a lot of progress was being made. At the end of the first three months of investigations, Rick held a meeting to which Wayne Grotzer was invited. The task force was quite pleased with its internal organization and the results achieved to date. They really knew the various processes used in the company as a product progressed from Marketing down to Distribution. They had found many examples of duplication of effort in different shops, of information being misinterpreted, of information not flowing to the right person at the right time, of people doing jobs that were completely unnecessary. Even without introducing automation, productivity could be increased by making improvements to the current manual process. Anna Bellinaso and Ed Falcon had completely mastered the financial and productivity aspects of automation technology. They had even developed a special procedure to compare the costs and benefits of different technologies. They delighted in analyzing the productivity boasts in advertisements and often found that the results were calculated incorrectly or were derived from data that appeared very suspect. Paul Valdas, Arthur McIlroy, and Malcolm Byers were working well as a team. Paul specialized in CAD, Arthur in NC, and Malcolm in robotics. At the meeting, the task force informed Wayne Grotzer that one of their major concerns was to find a solution in which existing designs could be easily retrieved for reuse on a new product, and in which data could be passed by computer from one application to another. They were surprised that in some systems it was difficult to recall an existing design. They also felt it was wrong that data from some CAD systems had to be drawn out on paper and then transferred to a part programmer to enter into another computer's NC system. They felt that what was needed was a system where the design data in the computer could be easily recalled and could also be used to directly produce manufacturing data.

That evening, on their commute home, Wayne and Rick discussed the next three month's work for the task force. They agreed that the first step would be to produce a detailed list of the major functions needed for the CAD system and for the NC cutting tools. After that they would call in the vendors, carry out some benchmarks, preferably at user sites, and then try to get down to a short list of three systems to be evaluated in more detail.

It took about two weeks to draw up the detailed list of required functions and decide upon their relative importance. Malcolm Byers had been very impressed by one particular CAD system and emphasized the importance of various functions, hoping the task force would consider this system. The members of the task force, however, wanted to ensure that the final list was not biased toward any particular system. The three best NC cutting tools were soon identified, but it was a completely different state of affairs for the CAD system.

The task force was glad it had prepared its prioritized list of functions before the CAD system vendors arrived. It was very difficult to understand the capabilities of the CAD systems—it was doubtful if some of the salesmen really understood them. The task force was told that missing functions would be available for "the next version," or "next year," or "by the time you need them." The computer that one system ran on was about to be upgraded so "even bigger productivity increases would result." On another system, "a new graphics screen would result in significantly faster throughput." Different members of the task force were invited to demonstrations of "next versions" in Florida, in California, and even in France. On one such visit, after much good wine had been consumed, a task force member told one vendor's marketing director that his proposal was 10% higher than his competitor's. Within a week, an apology and a new proposal were received from the marketing director—due to a fault in the word processing system, the first proposal had been overpriced by 15%. Another vendor showed results of a highly confidential survey carried out by a major automobile manufacturer showing that their system "produced drawings twice as fast as any other."

The task force had some problems getting the vendors to carry out standard benchmarks. One benchmark involved the design of a new piece of furniture and preparation of the corresponding NC tapes. The task force saw the benchmark as being a standard against which the different systems could be compared, and therefore wanted each vendor to carry out exactly the same work. If all the vendors did not carry out the same work then the benchmark lost much of its value. One CAD system vendor claimed not to have the time to do all the benchmark but would just do the most difficult part. Another vendor had time to do the design but not to prepare the NC tapes. A third vendor agreed to do all the benchmark, but warned that the results should not be given too much significance as the person who would do the benchmark was a newcomer. A fourth vendor could do all the benchmark—but in two months' time. In the meantime, it was suggested that the task force come and see the result of a benchmark that had been carried out for someone else. Another vendor took the task force for a one day visit to a user site. The morning was taken up by one of the users explaining how good the system was. Lunch was taken at one of the best restaurants in town and finished at about 2:35. They actually got to the CAD terminal just after three o'clock and were given a one hour demonstration of the systems capabilities after which a designer started to carry out the benchmark. Ten minutes later, he sprang from his chair, apologized and rushed off saying that he had to go to a meeting.

In spite of all the difficulties, the task force eventually managed to select three systems that appeared to come as near as possible to the list of required functions. Two of the systems were CAD/CAM systems (i.e., data from the design phase could be used directly as input to the manufacturing phase). The other selection was made up of a CAD system and a NC programming system. Although these were marketed by different companies, they ran on the same computer, and a special program had been developed to transfer data between them. The major advantage of this selection was that the functions in the NC programming system were more powerful than those available in the CAD/CAM systems.

In the last month before the six month deadline was reached, more thorough technical benchmarks were carried out on the short-listed systems. An attempt was made to simulate how each of the proposed NC cutting tools would be used with each of the proposed CAD systems and NC programming languages. Financial benchmarks were carried out to identify all the five year costs and benefits resulting from choice of the different systems. Organizational benchmarks were carried out to show how each system would be managed and implemented.

The day before the six months were up, Wayne Grotzer met with the task force and received its report. Rick Nelson gave a 30 minute presentation which he started by recalling the major objectives of the company and the resources available. He went on to describe the needs of the company as seen by the task force, and outlined the types of solution that had been envisaged before deciding on a short-list. Rick then showed the results of the detailed analysis of the short-listed systems, and gave the reasons for proposing one particular solution.

Wayne Grotzer asked if the task force's proposal was unanimous. Rick asked each of them to reply, and each supported the recommendations of the report. Wayne thanked them all for their hard work and committment, and thanked Rick for leading the task force.

Fritz Froehlicher took Wayne and Heidi to dinner that evening. Fritz raised his glass and said: "Here's to my son-in-law. Thanks to him Golden Triangle International Furniture Inc. is one of the few companies in the world that has managed to select the most suitable CAD/CAM system in less than one year." As Wayne looked lost for words, Fritz went on: "And they tell me that's the easy part. It only starts getting difficult when you get to the implementation stage."

Before looking at the implementation, let's leave Pittsburgh and take another look at the selection process.

3.2 PRODUCTIVITY AND FINANCE

The terms "productivity" and "productivity ratio" have a variety
of definitions. They are often used indiscriminately to express in
some way the ratio between the result of a productive activity and
the total effort put into the activity (i.e., the ratio of output to
input). CAD/CAM system vendors sometimes use productivity to
mean the ratio of the number of drawings produced with the
CAD/CAM system to the number of drawings produced without
the system. This is an unusual use of the term productivity as it
only considers the ratio of two output quantities without taking
account of the change in input effort. The input effort ratio
should also be considered since the input effort has been increased
by the cost of buying and implementing a CAD/CAM system.
Since the terms productivity and productivity ratio are so vague,
their use in a particular case has little meaning unless it is defined.
When they are used to describe a CAD/CAM system it is necessary
to define for example exactly which application is being carried
out with the system, the experience and ability of the people
carrying out the application, and how much of the application can
be carried out using the system. It is necessary to define which
application is being carried out with the system since some applications are currently more amenable to CAD/CAM than others,
for example, electronic circuit design with its restricted set of
symbols and characters will be expected to give a higher productivity ratio than design of complex mechanical parts such as
hydraulic valves. It is important to know the experience levels of
the people using the system. It may well be that an inexperienced
user of a CAD/CAM system takes longer to carry out a task with
the system, than an experienced craftsman would take to do it
without the system. Does this imply that the system should never
be used?

The more factors that are considered, the more difficult it
becomes to attach any real meaning to the terms of productivity
and productivity ratio. Consider the case of an application that
required 100 hours of work by a skilled craftsman, a CAD/CAM
system that has a productivity ratio of 5 to 1 for this type of

application, and another CAD/CAM system that has a productivity ratio of 2 to 1. If the first system can only be handled by a specialist who has used it for three years, whereas the second can be used by anyone after two weeks training, which system is more suitable? Or assume that the first system can only be used for four hours work on the application, whereas the second can be used for ten hours. Which system is more suitable? What does productivity ratio really mean for these systems? If it refers to time, what conclusion would one draw if it turned out that the first system costs five times as much as the second? If it refers to cost, is the cost considered the sum of the cost of the system per hour and the direct labor cost per hour, or has the cost of training and installation also been taken into account?

It is apparent that calculation of direct productivity ratios (i.e., those where it appears that two measurable quantities are being compared) becomes very difficult if an attempt is made to be really precise. When indirect productivity gains (e.g., more accurate design, improved morale, or more easily enforced standardization) are considered, it becomes apparent that a lot of subjective judgements go into calculating the productivity ratio. In the previous example, a CAD/CAM system had a productivity ratio of 5 to 1. The vendor might have meant by this figure that a task that took 100 hours without the system would take 20 hours with the system. Indirect productivity increases in this case might arise from the system being able to detect design errors that would have gone undetected otherwise, or from use of the system resulting in a reduction of time in programming NC quality control machines. Apart from such technical indirect productivity increases, there can also be business productivity increases (e.g., use of the system could result in winning a $1 million order that would otherwise have been lost).

· When trying to carry out a financial evaluation of a CAD/CAM system it is preferable to start with some figures that can be understood and agreed by all, rather than using figures that will always be open to dispute. Rather than trying to evaluate productivity ratios, it is better to start with the costs of the system (after all there are enough of them). It is best not to include the major costs arising from integrating CAD/CAM with other islands of automation. These costs should be considered in the CIM budget.

Assume that the system will be amortized over five years on a straight-line basis. Costs should be calculated for each of these five years and will include:

The initial purchase of hardware and software

Further purchases of hardware and software in later years (successful systems will expand)

Shipping and installation costs

System maintenance costs

The cost of training users throughout the five years

The cost of unproductive time (e.g., while some users are being trained, others may not be able to do their normal work)

The cost of running the system (management and operations)

The cost of selecting the system

Development costs for symbol and part libraries

Costs resulting from modifying existing manual and automated procedures to fit the system

Costs of various supplies and accessories not included in the major hardware and software costs (e.g., magnetic tapes)

The cost of travelling to visit other users and of going to user group meetings

Exact analysis of the CAD/CAM investment requires a very detailed knowledge of a company's financial situation. Even factors such as the geographical location of the company will influence the results. However, an attempt will be made to show the type of costs that can occur. It should be noted that initial hardware and software costs can range from $1000 to $1,000,000 and it is therefore difficult to give really typical values. The following costs could be incurred over five years for a system of initial cost $500,000 and a maintenance cost of 1% of the value of installed system hardware and software per month. It is assumed that four workstations are purchased initially with one extra workstation being bought in each of the following years. The costs do not relate to any particular system, but are fairly typical and show clearly that the cost of the initial system is not the only cost involved.

Initial system purchase	$ 500,000
Further hardware + software purchase	$ 400,000
Shipping and installation	$ 20,000
Maintenance	$ 500,000
Training	$ 300,000
Unproductive time	$ 50,000
Running costs	$ 400,000
System selection	$ 90,000
Library development costs	$ 50,000
Modification costs	$ 50,000
Supplies and accessories	$ 40,000
Travel and meetings	$ 100,000
Total	**$2,500,000**

It will be seen that the initial purchase price of a CAD/CAM system is just the tip of the iceberg. In the above example, it represented just 20% of the five year costs. The costs can be divided into the two major categories of equipment costs and personnel costs. Within the equipment category are included the hardware (computer, workstations, plotters, etc.) and software costs, their associated maintenance costs, and the costs of shipping, installation, supplies, and accessories. The total equipment cost will be very much a function of the initial cost of the system. The cost of maintenance varies from system to system. For some low-price systems there is no maintenance cost at all, but a typical cost is 1% of the installed system hardware and software value per month.

It is important for a company to define correctly the system that will be needed. It may be that a company's requirements can be met better by a system costing $100,000 as by one costing twice as much. The wrong choice can be disastrous in more ways than one. Not only would the company be wasting money, but purchase of the more expensive system could result in the company finding that "CAD/CAM will never be productive," whereas the cheaper system would result in major savings. Factors such as currency variations, the size of a vendor's marketing force and discounts offered by vendors to enter a particular industry or area can cause the cost of a system to get out of step with that of competitive systems. In a market where there are several hundred systems on offer, it certainly pays to investigate a variety of solutions.

Within the category of personnel costs are training costs, running costs, costs of travel and meetings, system selection costs, the cost of developing libraries, and the cost of integrating the system with existing procedures. The last two costs mentioned will vary greatly from one company to another. It is important that they be identified at the time of system selection. The other personnel category costs are basically a function of the number of people that will use the system. For eight workstations used 10 hours a day, there may well be 15–30 users (depending on the type of applications) and an operations and management support staff of about 4 people. Each user has to be trained, and the more users there are, the more management and support staff are needed. The more users there are, the more meetings and exhibitions they will go to. Almost independently of the type of CAD/CAM system (whether it is made up of eight workstations attached to a superminicomputer or eight individual engineering workstations), the costs in this category will be similar, about $5000 to $10,000 per user per year.

The figures given do not represent the actual costs of a particular system in a particular company. They only serve to illustrate the type of costs that are associated with CAD/CAM.

Some of the costs, such as the cost of hardware, are very visible—others, such as the cost of system selection are generally not seen. An attempt can be made to divide the costs up into visible costs and invisible costs. Often only the visible costs appear in a cost justification of CAD/CAM.

Visible costs
    Initial system purchase $ 500,000

Visible costs	
Initial system purchase	$ 500,000
Further hardware & software purchase	$ 400,000
Maintenance	$ 500,000
Total of visible costs	$1,400,000
Invisible costs	
Shipping and installation	$ 20,000
Training	$ 300,000
Unproductive time	$ 50,000
Running costs	$ 400,000
System selection	$ 90,000
Library development costs	$ 50,000
Modification costs	$ 50,000
Supplies and accessories	$ 40,000
Travel and meetings	$ 100,000
Total of invisible costs	$1,100,000

It can be seen that about 45% of the costs are invisible. Hiding the costs like this can be useful when cost justifying the CAD/CAM system, but someone will have to pay for it in later years.

Another way to look at the costs is to divide them between visible first-year costs and other costs.

Visible first year costs	
Initial system purchase	$ 500,000
First-year maintenance	$ 60,000
Total of visible first year costs	$ 560,000
Other costs	
Further hardware and software purchase	$ 400,000
Shipping and installation	$ 20,000
Other maintenance	$ 440,000
Training	$ 300,000
Unproductive time	$ 50,000
Running costs	$ 400,000
System selection	$ 90,000

Library development costs	$ 50,000
Modification costs	$ 50,000
Supplies and accessories	$ 40,000
Travel and meetings	$ 100,000
Total of other costs	$1,940,000

Here more than 75% of the costs have been hidden. It should be much easier to justify spending $560,000 than spending $2,500,000. However, if top management is told that the benefits of CAD/CAM can be had at a cost of $560,000, there may be some reluctance to paying out another $485,000 in each of the following four years.

The reason for buying a CAD/CAM system is that its use should result in direct and indirect benefits which will outweigh the costs. In the above example, initial system purchase may include four workstations, with one extra workstation being bought in the second, third, fourth, and fifth years. Assume that the company aims to use each workstation productively for 10 hours per day (this excludes unproductive use such as maintenance, downtime, idle time, and training which taken together will probably average two to three hours per day). If the system can be used 5 days a week and 48 weeks a year, then 2400 hours can be worked at each workstation each year. (Although the actual number of hours worked will vary greatly from company to company, in almost all cases the percentage of time during which such an expensive piece of equipment is gainfully used is surprisingly low. In a 365 day year there are 8760 hours, thus 2400 hours use in a year only represents 27% productive use.)

In the first year, the four workstations will clock up 9600 hours use, and over the 5 years the overall total for all the screens will be 72,000 hours. To recoup the total cost of $2,500,000 over five years requires recouping $34.72 per productive workstation hour. If the person who uses the workstation usually costs $17 per hour then, to break even, during 1 hour at the screen (at a cost of $17 + $34 = $51) the user must do the equivalent of 3 hours manual work (at a cost of 3 x $17 = $51). Another way of looking at this, on the assumption that the user continues to cost $17 per hour, is that during one hour, in addition to carrying out the same work as before, the user must make direct or indirect savings to the value of $34 (the cost of the workstation for that hour). If it is estimated that an indirect profit of $1,500,000 will arise from use of the system over five years, then only $1,000,000 has to be recouped directly (i.e., about $14 per hour at the workstation).

So far the figures, although they are only approximate, are relatively straightforward. However, in the list of costs, all costs incurred throughout the five-year period were included at the price levels ruling at the beginning of the five-year period (i.e., it was assumed that money has a constant value over time). As is especially apparent in times of high inflation and high interest rates, money does not have a constant value and the change in its value should be reflected in the calculations.

The basic payback method assumes that money does have a constant value. In the above case, if it can be shown that the system will result in benefits of (or pay back) $800,000 per year, the payback period would be just over three years ($2,500,000 divided by $800,000).

The change in the value of money makes the results of the calculations more difficult to interpret. Consider the alternatives. First, the company leaves $2,500,000 in a bank account for five years at 11% interest per annum. At the end of five years, the money will have grown to more than $4,000,000. On the other hand, if the company hands over $2,500,000 to a CAD/CAM vendor at the beginning of the five year period and receives a benefit of $800,000 per year then, at the end of the five year period the total benefit will be $4,000,000. Whereas the payback method appeared to support the investment in CAD/CAM, this calculation appears to show that the company would do better to put its money in the bank. (If the calculation were to be as lifelike as possible it would be necessary to take into account that the investment would not all be made at the same time, that the benefit from the system would not be the same each year but would be small to start with and increase with time, etc.) To take into account the changing value of money with time, the net present value method of calculation is used, but an explanation of this is beyond the scope of this book.

It is so easy to play around with the figures and produce all sorts of justifications to invest, or not to invest, or to wait a little longer before investing, that it may well be considered that these calculations are completely useless. This is not quite true, since they can be useful in comparing different options and different implementation schedules.

They cannot however be expected to decide whether or not a company should invest in CAD/CAM. The data that goes in to the calculations is available to managers. Management has a lot more information on the company, the product, the market, etc., that does not go into the calculation. Management, not magic formulae, must decide whether or not CAD/CAM is necessary.

When the magic formula does not work (i.e., when the right tool is not available for management), there appears inevitably an unwillingness to decide. The seventeenth century craftsman did not have this problem because he understood his business. The problems involved in deciding on how to invest in CAD/CAM are related to the departmentalization of the business, and the lack of an overall understanding of the business by many managers. In the short term, they can be overcome by taking decisions at a level higher than the department level (i.e., by top management).

3.3 SELECTION OF A CAD/CAM SYSTEM

Before selecting a CAD/CAM system, a company must be sure that it really needs one. A CAD/CAM system is not a miraculous solution that will solve all the company's problems—it is just a tool that, if used properly can improve the company's overall performance. Thus, before selecting a particular CAD/CAM system, the first thing to do is ensure that this is the most effective automation technology to invest in.

3.3.1 Selecting the Most Appropriate Automation Technology

In deciding on priorities for the implementation of automation technologies, a company-wide plan has to be drawn up taking into account the company's overall objectives and resources and the potentially available technologies. The market strategy of the company and its overall goals have to be formalized and analyzed in order to detect the operational needs. In particular, care must be taken to identify the characteristics of the product that, in the global sense, meet customer requirements such as cost, quality, physical characteristics, delivery time, and so on. These characteristics should then be mapped out to show relative strengths and weaknesses compared to competitors and to market requirements. It may become apparent for example, that although a competitor produces a more compact and higher quality product, these factors are surplus to market requirements. It may also become apparent that what is really required by the market is only a slight increase in quality, but a major reduction in delivery time. The automation techniques would therefore be chosen to meet these goals.

All available resources have to be examined, in order to define the base from which the automation technology implementation plan will be built. Obviously, financial resources need to be detailed, but so do existing systems and equipment, the human resources (skills and numbers), the information flow in the company, and the interfaces between different parts of the company.

The existing resources should then be confronted with the strategic marketing objectives in order to find out discrepancies, and thus the relative importance of potential automation technologies. This comparison should be made not only for the current situation, but also for the expected situation over the following 3 to 5 years. In this way, the timing of implementation steps can be determined and coherence given to the overall automation plan.

In the case of a company for which the major marketing goals require increased quality and reduced delivery time, there would be many automation technologies available. CAD/CAM could be used to improve product quality, but so could NC measuring machines, or, going back to the beginning of the product process, so could machines for automatically analyzing the quality of the raw materials. To improve delivery time, CAD/CAM may also be a potential technology. Others could be a production planning and control system or an automated material handling system.

Once all potential automation technologies have been identified, they should be evaluated against a set of criteria weighted in accordance with the company's strategic goals. If CAD/CAM is shown to be a suitable technology, then the next step is to increase CAD/CAM knowledge in the company to a level at which a reasoned choice of system can be made. After that, a full evaluation will be carried out on a limited number of CAD/CAM solutions. This should lead to selection of the most suitable solution and to its installation and use. The three phases of selection are shown in Table 3.1. They can also be applied to the selection of other advanced manufacturing technologies. The exact resources that a company commits to the selection process will depend on company size and other factors.

2.3.2 CAD/CAM Information Build Up

Once management has decided that an investment in CAD/CAM is required, it is necessary to move as quickly as possible to a position in which specific decisions can be taken. These decisions will answer questions such as: Which CAD/CAM system? How many workstations? Who will use it? How will it be used? On average, it will take about six months to build up CAD/CAM knowledge to a sufficiently high level to be able to make rational choices. This may seem a long time, but it must be realized that successful implementation of CAD/CAM takes a lot of time, effort, and management support. In many cases, about two years elapse between taking the initial decision to invest in CAD/CAM, and actually making some productivity gains. In the absence of proactive management, it may well take a further one or two years for the originally expected productivity gains to be attained. Once implementation starts, not only will people be trained and become efficient with the system, and data be continually entered into the system thus increasing the amount and value of information stored and retrievable, but also the company structure should change to allow information to flow more productively. In the same way

Table 3.1 The Three Phases of CAD/CAM System Selection

Phase 1: Information build up				
Management requirements	Existing engineering and manufacturing processes	Applications and skills profiles	Organizational aspects	Knowledge of CAD/CAM

Phase 2: Full evaluation of a limited number of systems				
Technical benchmark	Economic benchmark	System supplier benchmark	Organizational strategy	Simulation of initial system use

Phase 3: Management decision

that use of computers has resulted in change in financial organizations, and in the office, so it will change the existing structures in the engineering and manufacturing process.

The Task Force

Top management cannot be continuously involved in the CAD/CAM information build-up phase, and should delegate this activity to a specially constituted task force. The role of top management will be to direct, control and support the task force, and eventually to take the decision as to which CAD/CAM system (if any) should be purchased.

The task force should be led by a senior manager with an engineering and/or manufacturing background. Whereas the task force leader's role should be a full-time one, other team members should only have part-time involvement (about 1 day per week each). One member of the team should be assigned the role of task force secretary. This person should be a dynamic young manager with good administrative experience, and will act as a focal point for all incoming information and for arranging meetings and visits. The task force will be completed by another three to six people representing balanced, different views of the company. Someone from finance should be involved, as should a member of the EDP department (if there is one). Engineering and manufacturing should each be represented by one or two people. The type of person selected could be a forward-looking, senior, experienced manager or a potential system user (such as a part programmer, or an open-minded designer with an interest in computers). If already designated, the future CAD/CAM manager should be included in the task force, with the dual aims of gaining knowledge and increasing commitment to the final choice.

The task force must be given enough power to carry out its functions (other staff may resent its members "interfering" in their personal empire), and told when to report. For example, brief interim reports may be scheduled for presentation after two months and four months, with the final detailed report due after six months. It must be made very clear to the task force that its only objective at this stage is to raise the level of knowledge to a level at which a choice of CAD/CAM system can be made. The objective is not, at this point, to make that choice. This is an important point, as some task force members may be over-attracted to one particular system at an early stage, and as a result, not collect full and neutral information on other systems.

During the information build-up phase, the task force should look in detail at the following four areas.

Management requirements
The company's engineering and manufacturing processes
The company's applications and skills profiles
Knowledge of CAD/CAM systems

It will be seen that three of these areas relate specifically to the company, and only one to CAD/CAM systems themselves. This reinforces the point that unless a lot of effort has gone into understanding what the CAD/CAM system should do in the company, it will be easy to select a system, but the system selected may well be the wrong one.

Management Requirements

The final choice of system should depend very heavily on top management requirements. These should be defined by a management committee composed of selected company directors and top managers. This committee has a key role to play throughout the CAD/CAM selection, implementation, and utilization phases. Only top management really knows why the company should invest in CAD/CAM. Is it a short-term decision based on the need to improve the direct productivity of a single application? Is it a medium-term decision linked to the expected implementation of a flexible manufacturing system? Is it part of a long-term plan aimed at automating all the operations of the company? What are management priorities on the benefits that should arise from use of the CAD/CAM system—reduced subcontracting, improved quality, reduced lead-time, use of existing parts in new designs, faster and more accurate quotations? What type of return does management expect to get from the system, how much can be invested? Only top management can know the answers to these questions at the company level. Perhaps management will not want to give away such information, but at least by asking the questions, the task force may lead management to look for the answers and therefore be in a better position to make the final choice of system. If management is not aware of how it expects CAD/CAM to impact overall company performance, the impact will probably be negligible.

Apart from its role as a provider of information on overall requirements, management has a very important role to play in supporting the task force. The general inertia of the company will have to be overcome in implementing a radical new technology such as CAD/CAM and this cannot be done without management commitment. Top management should make it clear to all concerned that a technology that is potentially very beneficial to the whole company is under consideration, and that the task force has complete management support. As the selection process is carried out, top management must be preparing for the changes in organizational structure and corporate perspective required for successful CAD/CAM.

The Engineering and Manufacturing Process

The majority of CAD/CAM systems marketed today offer a standard "black-box" solution in which the system cannot be tailored to fit the requirements of a particular company. Since all companies are different, and have different requirements, the major problem in system selection is for the company to find the black box that comes as near as possible to meeting its requirements. It is therefore very important that the task force understands how the company's design and manufacturing activities appear from a CAD/CAM point of view.

These activities all use information (in some form of data) and the task force must understand how information is created and manipulated as a product goes from conceptual design through to production. This process should be looked at as a whole—not as a succession of steps separated by the existing departmental structure of the company. A typical product path might include marketing, design, analysis, drafting, process planning, tooling, NC programming, machining, assembly, quality control, and maintenance. All these processes use data (or information) in some way. The computer department may already be handling some of the data, but will probably not be completely aware of its global, interconnecting nature.

The product path may be thought of as being made up of a series of activities. Each activity consists of input of data and commands, a process (e.g., drafting or NC programming), and output of data. As the product path is investigated, it will become apparent how the product is produced today (i.e., how many people work in each activity, how they manipulate and communicate data and so on). The task force must dissociate the underlying processes from the way in which the product is produced today. Whereas the underlying processes may not change with CAD/CAM, the way in which they are carried out probably will change.

Applications and Skills Profiles

As well as looking at the product path from the information flow point of view, it is also necessary to investigate the individual processes in detail. Major areas where CAD/CAM is usable can be identified. A process-related applications profile and a skills profile of the users can be defined for each process. These can then be combined to give company-wide applications and skills profiles which can be matched against the capabilities and requirements of individual CAD/CAM systems. In drawing up the applications profile it is important to assess how much of the work involves creation of new parts and how much involves modification of existing parts. There are CAD/CAM systems available that are much more efficient in creative work than for modifications, and vice versa. The complexity and shape of parts produced are other factors that must be evaluated. If many parts have similar shapes or follow similar manufacturing processes, it may be possible to group them into families of parts and make savings by applying parameterization or group technology techniques.

The applications profile should contain precise numbers on, for example, the number of drawings produced in a given time, the life-time of a drawing, the number of part programs that have been written, the average number of times a part program is modified. The profile should also contain information on key issues such as standards used within each activity and control procedures (e.g., for drawing release). Whereas the applications profile will reflect the product and processes, the skills profile will reflect the people who are potential users of the system. This is necessary information, since it would be unreasonable to buy a CAD/CAM system that no one will use because it is too difficult to handle, or conversely is too pedestrian. The skills profile will also show whether or not existing staff can support the CAD/CAM system, enhance its capabilities with additional programs and interface it to existing systems in the company.

Again, it would be unreasonable to buy a system that could not be interfaced to other systems, since one of the major benefits of CAD/CAM arises from the reuse of data in several applications. Existing systems (computer-based, automated, or manual) that must interface directly with the CAD/CAM system should be investigated in detail to see exactly how they will work alongside the CAD/CAM system. Some thought should also be given to the relationship of the system to other automated technologies that may be introduced in the future. As a final point concerning the interface with other systems, some attention should be paid to CAD/CAM systems used by suppliers and partners. If there is a high volume of data transfer between different companies (or plants), the capability for correctly and completely transferring such data may be a very important criterion for system selection.

To avoid future recriminations, the applications and skills profiles should be validated with the managers concerned.

Learning About CAD/CAM Systems

The task force must be very careful to learn about CAD/CAM systems without making premature and ill-considered judgements. There are several ways of learning about systems:

By reading magazines, journals, and books
By attending exhibitions and conferences

By visiting CAD/CAM users (preferably companies with similar applications and skills profiles)

By contacting CAD/CAM suppliers (this should only be done once a good basic knowledge of CAD/CAM has been attained. As a first step, each supplier should be requested to send full descriptive documentation of their systems, detailing the particular functions that are available. The suppliers will be only too happy to demonstrate their wares, but there is little to be gained from attending such demonstrations without a good basic knowledge of CAD/CAM).

The first two ways of learning listed above are impartial but may be time consuming and expensive. The last two ways will almost certainly be biased towards one system or another, but have the advantage of rapidly showing the merits of a particular system—and probably hearing what is bad about other systems. Another alternative, which can lead to savings in both time and money, is to use a well-qualified, objective, external specialist such as a consultant. It is to be expected that the management consultant have not only specific CAD/CAM knowledge, but also be able to take a broad view of the company and identify the overall role of CAD/CAM.

The best solution will probably be a mixture of the different alternatives and after a few months, the task force should be well aware of the advantages and disadvantages of the different types and components of CAD/CAM systems:

Turnkey system, service bureau or software package

CAD, CAM, CAD/CAM, CAE, etc.

Mainframe based, superminicomputer based, minicomputer based, engineering workstations based, or microcomputer based

Vector refresh, raster or storage graphic screens

Centralized or distributed

2D, 2½D, or 3D geometry modelling

Wireframe, surface, or solid modelling for 3D geometry

Design, drafting, analysis, and machining possibilities

Data storage and retrieval capabilities

The links between CAD, FEA, and NC
Etc.

If task force members are not familiar with the terms used
above, then they are not ready to progress to the next phase of
the selection process.

The four activities of CAD/CAM information build up (concerning management requirements, the engineering and manufacturing
process, the applications and skill profiles, and CAD/CAM system
knowledge) should be carried out simultaneously. No doubt there
will be overlap and interplay between them. However, something
learned on one activity will often be of use in increasing understanding in another area.

It is important to remember that during this phase, the major
requirement is not to select (and defend) one particular system,
but to build up a global picture of the processes in the company
that CAD/CAM can be applied to, and the ways in which CAD/
CAM systems in general can meet these requirements.

3.3.3 Full Evaluation of CAD/CAM Solutions

At the beginning of this phase, the task force should draw up a list
of CAD/CAM systems available and a prioritized list of CAD/CAM
system functions required by the company. The company applications and skills profiles will be used alongside these lists as the task
force aims to reduce the list of available CAD/CAM systems down
to the four or five most appropriate systems. The overall CAD/
CAM knowledge that the task force has gained should allow some
systems that are obviously unsuitable to be eliminated from the
list. Knowledge of management requirements and available
resources should serve to eliminate other systems. Each of the
remaining four or five solutions on the list should now be thoroughly tested to see if it really would be suitable. The final tests
will be carried out in five parts:

A technical benchmark
An economic benchmark

A system supplier benchmark

Definition of the organizational strategy associated with each system

Simulation of initial use of each system

The Technical Benchmark

At the beginning of this phase, the task force will have available its prioritized list of required CAD/CAM functions. It should now define a simple part, a part of average difficulty, and a difficult part to be tested on each system. The parts should be typical of the workload of the company, and the tests should include all functions required, not only in drafting but also in design, analysis, and machining. The task force secretary should then contact each of the four or five potential system suppliers requesting them to state in writing the capabilities of their system to carry out the functions on the prioritized list. When the replies have been received, the task force secretary should make available the tests to the system suppliers and request that each supplier demonstrates, over one or two days, how its system would carry out the test.

The technical benchmark should ensure that a system will be capable of carrying out the required functions. It should be broad ranging and cover in depth areas such as:

Geometric modelling

Data management

Parameterization functions

Project control functions

Hardware

Application areas assisted

User interface

Assembly functions

Drawing management

Potential interfaces to other systems

The technical benchmark is meant to be a test of the required functions of the CAD/CAM system, not of an individual operator's knowledge of the entire system or ability to handle a particular company's applications. It is, therefore, reasonable to give a supplier some time to prepare the demonstration. Even at the demonstration, it is important not to take too much account of the overall time taken to carry out the tests. It is more important to see which functions are used, how they are used together, and how the system responds to individual functions. If system response time is too long, users will soon become frustrated and will not want to work with the system. (It is important to see how many other users are using the system during the benchmark, as response time is greatly affected by the number of users.)

Some functions of a CAD/CAM system are very difficult to benchmark during a one or two day test. The experience of an independent consultant can be invaluable in assessing functions such as data management capabilities.

The Economic Benchmark

Each system supplier should be asked to make a full proposal including price, delivery time, training time and location, whether or not training is charged separately, possibilities to upgrade the system, availability of technical assistance, support staff required, maintenance costs, installation costs, and so on. Each proposal should be fully costed, bearing in mind all hidden costs (ongoing training, management, operation, etc.).

The applications profile and the skills profile should be used along with the experience of the task force to estimate the potential benefit (business and technical, direct and indirect) of the system to the company. Once the investment and running costs have been evaluated, total systems cost (adjusted if necessary for sales of replaced equipment and taxes), and quantified benefits arising from use of the system, then form the basis for net present value (NPV) and internal rate of return (IRR) calculations.

The System Supplier Benchmark

Each system supplier should be asked to reply in writing to a set of business questions. What has the annual turnover and number of employees been over the previous five years? (In a market growing at about 30% per year, growth in turnover of 15% means a loss of market share.) How long has the company been in business? How many systems have they sold in the previous five years—to existing system users and to new users? Which other companies in the same industrial sector use the system? Which is the nearest user geographically? Where is the service engineer located, which spares are held, how long will it take to repair major system problems? Existing system users should be asked if the service and technical support is sufficient, and if it has been possible to interface other engineering and manufacturing systems to the system. Answers to the above questions must be treated intelligently as they may favor the largest suppliers on one hand, and on the other hand the suppliers that do not produce audited annual results (perhaps because they are minor divisions of huge conglomerates).

It is important to investigate the reliability of the hardware and software in the system. If the hardware is produced or modified by the system supplier it may be more unreliable and less easy to upgrade than if it comes directly from a leading computing equipment manufacturer. Can the system under review be upgraded without being physically replaced? If the supplier stops producing hardware, can the system software run on other hardware?

Another important point to consider is the future development of the system. An attempt should be made to obtain a copy of the development plan proposed two or three years ago and compare it with actual developments. Similarly it is useful to know if the original developers of the system have left the company. If so, it is important to find out the qualifications and industrial experience of those now responsible for system development.

Arrangements and costs for basic and advanced training of users and managers should be evaluated. Manuals should be inspected. They may have been written in such a way by the original developer of the system that they are meaningless to a user. It is also useful to know if there is a "user group" for the system. This may be a useful source of information before purchase, and a source of assistance after purchase. The way in which the supplier maintains contact with users after purchase is important to clarify. Some suppliers designate someone to be the contact person for each client, others try to have as little contact as possible with their clients.

The Organizational Strategy

CAD/CAM is an important strategic technology for the company's future and it is necessary to ensure that organizational issues are considered at an early stage. CAD/CAM can be managed from the top down, or from the bottom up. Neither choice is always correct or always incorrect. The choice may depend on variables such as the product or services that the company offers, or the type of CAD/CAM solution selected.

In a top-down organization, CAD/CAM is run from the top by someone with a title such as productivity director. Such a function is highly visible and tends to concentrate the organization's focus on CAD/CAM. It implies recognition of the fact that CAD/CAM and data are company wide and not to be confined within departmental walls. It also focuses power and forces recognition of power—the power to enforce changes.

By contrast, in a bottom-up organization, CAD/CAM can be run by a CAD/CAM steering committee made up of representatives of users in different parts of the company. Reflecting the needs and problems of users of the system, the steering committee can ensure that CAD/CAM develops in the way that the users require.

In both types of organization it is useful to set up a CAD/CAM user group so that the CAD/CAM system users can get together to exchange experience, discuss methods, help each other, and complain. The organizational entity that has day-to-day responsibility for CAD/CAM is the CAD/CAM team. This will be run by a CAD/CAM manager, assisted by a team whose size will be very much a function of system size, company size, and company requirements. The team will operate the system, archive and release data, encourage people to use the system, train new users, assist users with new applications, and develop reporting procedures. The CAD/CAM manager will be responsible for planning use of the system and installation of new hardware and software when necessary. The CAD/CAM manager will also have to report to top management on the progress being made toward attaining the system objectives as originally defined. Both the link between the CAD/CAM manager and top management, and reporting procedures to top management, must be clearly defined.

Other organizational questions that should be looked at when evaluating a system include the attitude of trade unions, the possibility of working flexitime or multiple shifts and how use of the system will be paid for. It is best to give unions maximum advance notice of the introduction of CAD/CAM, and to ensure that they and the entire work-force are given full information. Such a policy provides the basis for an agreement that is acceptable to both sides. One major request of the union may well be that no employees lose their jobs because of the introduction of CAD/CAM. The unions may well resist flexitime or multiple shifts, but from the company's point of view it is a waste of money to under-utilize the CAD/CAM system, and so some form of agreement is required. Different companies bill the use of CAD/CAM in different ways—sometimes system cost is regarded as part of the company's general overhead, in other cases it may be charged as a predefined cost to a project, or on an hour-by-hour usage basis.

Simulation of CAD/CAM Use

At the end of the phase of full evaluation of CAD/CAM solutions, the task force must go to top management and confidently propose not only which system should be selected, but also how it will be installed and used over the first year. Thus, in addition to the technical, system, economic, and organizational aspects to be examined for each solution, the task force must try to evaluate:

The size of the initial system
How, when, and where the system will be installed
How and where training will take place
Who will be trained first
Whether a parts library of existing parts should be developed
Which applications will be treated first
How the system will be used
How the system will be interfaced to other systems in the company

An attempt should also be made to see how each solution allows for growth over a three to five year period. Over a five year period, some companies have multiplied by five the number of terminals in use. The effect on a company of a 30 terminal CAD/CAM installation can be very different from that of a 6 terminal installation.

In many cases, management wants to start with a very small test installation of CAD/CAM. Such an installation often turns out to be relatively unsuccessful just because it is so small, and therefore offers little scope for increasing productivity through the reuse of data. On the other hand, it is not realistic to try to switch completely in a few months from a totally manual installation to a CAD/CAM only installation. Once the decision has been taken to invest in CAD/CAM, stepwise implementation should start as soon as possible.

3.3.4 Presentation of Choice of Solution

When the evaluation is complete, the task force should prepare a detailed written report recalling the major company objectives, outlining the major benefits expected, describing in detail the technical, economic, and system supplier benchmarks, describing the organizational, implementation, and utilization issues, and indicating possible future developments. The report should also define criteria for the final selection, which when weighted in accordance with strategic marketing and financial goals lead to a preference for a particular solution. A detailed list of the actions required to implement the solution, and a corresponding timetable should be given. A five page summary of the report should be prepared for the busy executive.

In a one hour meeting, the task force should orally present its findings to management. Care should be taken to show why some systems were not considered in the final evaluation, and also to bring out the essential differences between the systems evaluated in depth. The criteria for final selection and their weights should be explained in detail. Although the strategic analysis may point to one particular solution, it is for management to make the final decision.

4

GETTING STARTED IN CAD/CAM

Once a CAD/CAM solution has been chosen, the list of actions required to implement the solution and the corresponding timetable must be agreed. These then become the responsibility of the CAD/CAM manager. Major areas of concern at this stage include (Table 4.1):

Clarifying the position and role of CAD/CAM in the company
Preparing for system installation and use
CAD/CAM system installation
CAD/CAM system use

Table 4.1 Major Areas of Concern When Getting Started in
CAD/CAM

Clarification of the position and role of CAD/CAM
   Define the overall CAD/CAM implementation plan
   Define the way the system will be used
   Define the organizational structure

Preparation for system installation and use
   Plan initial system use
   Start large-scale training and education
   Plan for installation of the system
   Develop a company CAD/CAM manual
   Define standards
   Define operating, project, and data management procedures

Installation of the system
   Carry out installation
   Carry out acceptance tests

Initial use of the system
   Start using the system
   Refine operating, project, and data management procedures
   Monitor system use

Before looking at these areas in detail, let's look at a fictional
example where little concern was given to organizing the implementation of CAD/CAM.

4.1 MONON MODELS INC. GETS STARTED
IN CAD/CAM

Unfortunately it is not possible to describe how Golden Triangle
International Furniture Inc. of Pittsburgh started their CAD/CAM
activities. Although Fritz Froehlicher and Wayne Grotzer accepted
their task force's proposal, they decided to delay purchase of the
system until the following financial year. However, Fritz suggested
that it might be possible to give some details of the implementation of CAD/CAM by the equally fictitious Monon Models of

McKeesport, Pennsylvania. Claude Le Paysan had moved to this company when he was offered the job of EDP Director.

Monon Models had decided to computerize its operations and had put out an advertisement for an EDP director with particular responsibility for introducing CAD/CAM operations. On the basis of the deep and wide-ranging knowledge of CAD/CAM he had acquired at Golden Triangle and an impressive performance at his interview with the CEO, Claude Le Paysan had been named EDP Director of Monon Models. In the interview, Claude had stressed the minimum fivefold productivity gain that CAD/CAM would yield, the low start-up cost of CAD/CAM, and its one year payback period. This had impressed the CEO who was looking for an EDP director capable of really driving the company into a new automated epoch. He wanted the transition accomplished quickly and within rather strict budgetary limits. The last thing he needed was one of those so-called managers who is in reality a strange mix of memo writer, clerk, and chaser.

Claude quickly got down to work. He decided that the first implementation of CAD/CAM should be in the 15 person Model and Mould Shop. This shop was supervised by Tom Bruggstein and carried out both design and manufacturing operations.

Claude called some friends he knew in the computer industry and eventually found one who could supply him with a CAD system very quickly. This was claimed to be a particularly powerful system with advanced data base facilities that would be of great benefit to Monon. Another company in Pennsylvania was already using the system. The following day, Claude took Tom Bruggstein out to see it. Tom had never seen an interactive graphics system before and was really impressed. He was amazed how a drawing could be zoomed, and lines could be erased from a drawing. The system was used for civil engineering, and Claude and Tom were shown some really outstanding drawings that had been produced with it. One of the users said that some other companies used the system for NC programming. Claude was sure they had found the right system, but he realized he would need a 40% price reduction to stay within the budget, and someone to operate the system. He decided to hire Eddie Ludd as the operator.

Eddie did not have extensive CAD/CAM experience, but he was eager to learn. Claude called Eddie and told him that he would be hired as soon as the CAD/CAM system was installed. Claude, trying to work within the allotted budget, managed to obtain a 35% price reduction by not signing the maintenance contract and agreeing to drastically reduced training. The vendor knew that Monon Models wanted the system installed quickly, but could only deliver in three week's time and, to hold on to his potential client was prepared to reduce the price by a few more percentage points. Claude, of course, agreed to this, and further agreed to buy an out-of-date central processing unit (CPU), and only one workstation in order to have quick installation.

Claude wrote a memo to the CEO about the progress made toward implementing a CAD/CAM system, but he was beginning to become apprehensive about how serious Monon Models was about automating.

Before the system arrived, Claude selected an elderly, reliable mould maker named Jake Kanutza to oversee use of the system, and a young part programmer, Paul Kim, to be the first user of the system. Together with Eddie Ludd, the CAD/CAM system operator, they would form the heart of the new CAD/CAM Operations Team. Just before the system arrived, a local company was hired to build a new, glass-walled CAD/CAM operations room.

The room was not finished by the time the system was supposed to arrive but, fortunately, the system arrived a week late. However, the system's late arrival meant that Paul Kim, the first user of the system, had to cancel his vacation.

On the Monday morning that training should have started, the system was not working very well and crashed about every 30 minutes. Eventually the trainer, Joseph Zynsky, found that the wrong version had been supplied. He telephoned the New Jersey supplier and it was agreed that the correct version would be flown in on the first available flight. In the meantime, Zynsky gave Paul an explanation of the system on a blackboard. By Tuesday morning, the new version had arrived and by eleven o'clock it was running successfully. This was fortunate because just a few minutes later, Claude brought the CEO in to see the new system. Joe Zynsky gave a demonstration of the system to the CEO, who was very impressed and decided that he would show the system that afternoon to some potential clients from a domestic appliance company in the hope of clinching a deal. It was agreed that Zynsky would give the demonstration, and he was asked to prepare some parts on the system beforehand. Paul Kim watched Joe Zynsky working, but found he could hardly follow the actions of the trainer who seemed to be pressing buttons at random and moving the cursor round the screen at a bewildering rate. Paul had no understanding at all of the logic behind all these movements.

At about four o'clock that afternoon, Claude and the CEO came to show their system to the visitors. The CEO assured them that with the system, Monon Models would be able to produce the models required within two weeks. "That's fantastic" said the elder of the visitors, "we were expecting you to say two months." "Well," said the CEO, "automation has done a lot for us. This CAD/CAM system gives us a fivefold productivity increase. Now, shall we go and discuss the details of the rest of the contract. Why don't you leave some of your drawings here so that our people can start to work on them." As soon as the visitors had gone, Claude rushed down to the CAD/CAM room, but since it was 4:35 Paul and Joe Zynsky had already gone.

The next morning, Claude fixed a 9 o'clock meeting with Paul Kim, Joe Zynsky, Jake Kanutza, and Eddie Ludd. Claude arrived at the meeting and proceeded to tell the others how important it was that this first use of CAD/CAM be successful. He asked Joe Zynsky if Paul was capable of doing the work alone, but Paul said that he had not learned anything yet. It was decided that during the remaining three days set aside for training, Joe Zynsky would try to design the part for the domestic appliance company. Eddie Ludd requested some CAD/CAM operating training, but Claude said that this was less important and could be picked up from a manual. It was when Claude asked when the part could be machined that Joe Zynsky really began to worry. As far as he knew, the part-programming software had not been installed, and in addition, he did not know how to use it. Although it turned out that Claude had not ordered the NC software, an agreement was reached under which Dwayne Johnson, a trainer proficient in NC would bring it as soon as possible to Monon Models. In the meantime, Joe Zynsky started to design the part. The next day, Thursday, Dwayne Johnson arrived with the NC part-programming software. Unfortunately, it was not compatible with the version of the design software installed at Monon Models as the structure of the geometric data files was not the same. The work that had been done to date had to be abandoned, and so on Thursday afternoon, the design had to be restarted with a version of the design software that was compatible with the NC software. The two trainers could not agree between themselves which was the best way to carry out the design, and Paul Kim found it very difficult to understand what they were doing. By 4:30 on Friday afternoon, he was only too happy to go home and forget all about Monon Models and CAD/CAM.

The following Monday morning, Claude Le Paysan called a meeting of Paul Kim, Jake Kanutza, and Eddie Ludd. He explained to them how important CAD/CAM was to the company and that now it was up to them to make sure that CAD/CAM succeeded in practice. Jake Kanutza had been busy the previous week, and still had not seen the system. He expected to be very busy on other projects over the next few weeks, and Tom Bruggstein had suggested that he should not get too involved in CAD/CAM. Eddie Ludd pointed out that he had received no training at all so far, and Paul Kim said that he had received very little training as the two trainers had been so busy designing the part for their client. Claude told them "Everything must be finished by the end of next week. Come on, we need a plan. Paul, when can you finish it?" Paul said that he had not started yet and did not know when he would finish. "Listen Paul, your company needs you" said Claude. "Do your best please." Paul worked evenings and weekends to complete the job, and even managed to train another designer while he was working. This way, when Paul stopped working, the other designer could take over and continue making progress. Paul had difficulty in continuing the design started by the two trainers, and things became very complicated when the second designer began to change Paul's design. The system had very limited three-dimensional wireframe geometry modelling facilities and it was very difficult to be sure that the views designed really corresponded to the required part. (It was claimed that in the next release, this problem would disappear, since solid modelling would be available.)

After two weeks of hard work on the part, something had been designed and it was decided to program the tool path. An attempt was made to check the tool path on the screen, but as the zoom feature did not appear to work properly, it was not possible to get a really close-up view of the path. Claude decided that the tool path was probably good enough, and that the part should be machined in a light steel. If necessary, someone could hand finish the part over the weekend.

On the following Monday morning, the CEO told Claude that he had managed to pick up a couple more jobs for the system, continuing "They looked about as difficult as that one we've just done, so I quoted two weeks for each one."

Claude had calculated how much time had actually gone into the first job. Including all the evening and weekend work, it came to 424 hours, which meant 53 days. It had actually taken longer to do the job with the system than it would have done to do it without the system.

Claude Le Paysan saw no way in which the two new jobs could be done in the time span promised by the CEO. He wrote a memo to Tom Bruggstein asking for his help. When Tom received the memo he didn't understand it, he had only just heard that the first job had been successfully finished and that two more jobs were on the way. He decided to go down to the CAD/CAM room and talk things over with Paul Kim. "Hi Paul. Enjoy your weekend? See the Steelers?" "Hi Tom," replied Paul. "I just rested up this weekend. CAD/CAM burns me out." "You've done a great job Paul. Take it easy for a while. Have you seen this note from Claude Le Paysan?" As soon as Paul had read it, he walked over to a keyboard and typed in a few characters. Then he turned to Tom and said, "I'm quitting." Before Tom could say anything, Claude rushed in and said, "I've just had a call from the domestic appliance company.

They want us to modify the design by tomorrow evening. Can you do it, Paul?" "No way," replied Paul. "I just deleted that model we did for the domestic appliance company as there wasn't room for it on the disk. You'll have to get it back from the back-up tape." "What back-up tape?" said Eddie Ludd, the CAD/CAM operator, "no one told me to do a back-up." Claude's greatest fears had been realized. Cutting corners to stay within the limits imposed had proved disastrous. To avoid a similar fate, read on and see how CAD/CAM should be implemented.

4.2 CLARIFICATION OF THE ROLE AND POSITION OF CAD/CAM

During the CAD/CAM system selection phase, many possible solutions for the implementation of CAD/CAM will have been discussed. To remove any doubt as to which solution has been chosen and how it is to be implemented, the CAD/CAM manager should produce a report clarifying all these issues. Basically, this report needs only to recall those options proposed by the task force that were selected by management. It should include a description of the overall, long-term, company-wide CAD/CAM implementation plan. It should also define how the system is to be used, and describe the organizational structure that will be put in place for CAD/CAM. The report should be given as wide a distribution as possible so that all concerned know what is happening, and have sufficient information on which to base the decisions that must be taken.

It may also be useful to inform clients that CAD/CAM is to be implemented.

4.2.1. CAD/CAM Management Structures

Issues concerning the management structure of CAD/CAM operations should be resolved as early as possible, preferably before a system has been selected. It is clear that the users of the CAD/CAM system will continue to report to their usual hierarchical managers, and the CAD/CAM team members will report to the

CAD/CAM manager. The major problem is to decide to whom the CAD/CAM manager should report.

The CAD/CAM manager is one of the key people in the introduction and use of CAD/CAM, and has several important activities:

Recruiting CAD/CAM team members

Ensuring that suitable education and training is available for all concerned

Ensuring, through the CAD/CAM team, that an efficient day-to-day service is available

Ensuring that the information within the system is stored properly and can be efficiently transmitted to and accessed by users

Meeting with the users' managers to plan for future use of the system—deciding for which projects CAD/CAM should be used, scheduling resources for projects, deciding when a new project can start on the system, etc.

Calculating system use and productivity, and reporting these to top management in comparison with original plans

Making long-range plans and budgets

Meeting with the CAD/CAM user committee to deal with complaints and suggestions

Developing the CAD/CAM system control and management procedures concerning system security, backup, maintenance, recordkeeping and so on.

Liaising with the vendor and external user groups

In a pure top-down organization, the CAD/CAM manager would report directly to the productivity director. In a bottom-up organization the great danger is that if initially the users are only drawn from one department (e.g., design and drafting), the CAD/CAM manager will automatically be drawn from that department and will continue to report to the corresponding department manager. Problems will arise as soon as users in another department want to use CAD/CAM. The solution that they (and their department managers) often reject is to put themselves "under" a CAD/CAM manager (from another department) who reports to the manager of another department. Having rejected that solution, they will probably go on to buy their own CAD/CAM system and set up their own CAD/CAM organizational structure. This, of course, defeats one of the major aims of CAD/CAM, which is to improve the flow of information throughout the company. To avoid interdepartmental problems it is preferable for the CAD/CAM manager in a bottom-up organization to belong to and report to a "service" department.

The type of duties that the CAD/CAM team members will have to be trained for include:

System optimization (thus minimizing response times and maximizing system throughput)

System operations (such as ensuring that workstation use is efficiently scheduled, carrying out backups and long-term archiving, ordering paper and tapes, knowing when and how to contact the maintenance engineer)

System development (both interfacing the system to other systems in the company and extending the system to include special company-specific functions)

CAD/CAM training and education (although initial training and education will be given by external staff, later some of the load will have to be carried by the CAD/CAM team. In time, highly motivated "superusers" willing to train their colleagues will probably emerge)

System selection (as time goes by, the CAD/CAM installation will be expanded, or even changed. There will be a continuing need for the ability to select new hardware and software)

Internal sales and marketing (although the first users of the system may be very positive towards CAD/CAM, there will be a lot of potential users in the company who will prefer to stand back and keep everything manual. To ensure overall CAD/CAM success, it will be necessary to coax these people toward the workstations)

4.3 PREPARATION FOR SYSTEM INSTALLATION AND USE

Once the global picture of CAD/CAM implementation and use has been defined, the CAD/CAM manager must ensure that detailed planning for system installation and use is carried out. From the overall picture it should be clear to which applications CAD/CAM is to be applied initially.

4.3.1 Planning the Installation of the System

One of the most important activities concerning the installation of a CAD/CAM system is the detailed planning phase. Some knowledge of installation will have been gained during the phase of system selection, when an evaluation was made to see how each candidate system would be used over the first year. This knowledge must now be put into practice. Detailed planning and site development may take several months, thus as soon as a system has been selected, work should start on preparing for installation.

One basic problem of installation is to decide where to put all the equipment (the computer, disk units, magnetic tape units, workstations, system consoles, line printers, plotters, disks, tapes, and so on). The decision may well depend on the geographical layout of the company, on the location of the various users, on the limitations of the system, and on the company-wide CAD/CAM organization.

The equipment can be split into four categories:

1. Information archiving equipment, such as tapes and disks, crucial to projects and the company, which must be kept in dangerproof storage
2. Equipment which only CAD/CAM team operations staff need to access (e.g., the computer, the disk units, the magnetic tape units, a system console, storage space for disks and tapes)
3. Equipment which will generally be handled by the operations staff but may sometimes be accessed by the users (e.g., line printers and plotters)

4. Equipment which will be handled almost exclusively by the users (e.g., the workstations, local hard-copy devices, and perhaps a system console for controlling a cluster of local workstations)

For each equipment category, an evaluation should be made of the physical size of site (or sites) required and the corresponding necessary environmental conditions. The system supplier should be contacted for exact details of the size of the equipment, the power source requirements, and the environmental requirements.

The equipment, such as the computer, which only the CAD/CAM team will access, should be sited in a controlled-access, specially designed room. Temperature and humidity will have to be continuously monitored and controlled. A false floor under which the cabling can run, an air filtration unit to keep dust out, special power sources and common sense (no eating, no drinking, no smoking in the room) will increase the uptime of the system. The layout of the room should include not only the main items of equipment, but also space for storage (of disks, tapes, and so on), space for operators to move about in, space for maintenance staff to take everything apart, and some space for expansion of the system.

Concerning the equipment that will be shared between the CAD/CAM team operators and the users, much will depend on how many units will be installed. Often, the first unit of each type is installed in a room adjacent to the computer. Again, this will probably have to be a specially designed room with temperature and humidity controls. As the drawings of the company's new products will be produced here, controlled access is again necessary.

Whenever possible, the user equipment such as the workstations and local system consoles, should be grouped in clusters. This eases maintenance, control, communication between users, and training. However, it may not always be possible to group the workstations, for example, if just one is needed in a particular location and all the others in another location sited 200 yards away.

The workstation area is where the real action takes place. If it is unsuitable for human use, then the company cannot expect to achieve maximum productivity gains from CAD/CAM. It should be remembered that the users will be in continuous interaction with the system—this is difficult and requires concentration. Everything should be done to ensure that they can concentrate—the room should be quiet (shielded from the noise of the computer room and line printers), at a constant temperature (neither too hot nor too cold), free from static, well away from busy corridors, and out-of-bounds to all potentially distracting people. If necessary, a window can be put in one wall so that company management can show "its" CAD/CAM system to visitors. The level of background noise should be high enough so that brief discussions between two users will not distract the others, but low enough to allow uninterrupted concentration. It is essential that lighting in the workstation room be suitable. Storage screens in particular give dim images that cannot be worked with in normal daylight over long periods, thus it may be necessary to install curtains. Reflections in the workstation screen can be very annoying to the user, and lighting should be so installed as to reduce these.

One workstation in a cluster can be regarded as "special," in that it may sometimes be used in group training, or for maintenance or system work. Extra space should be planned around this terminal for training session equipment and a telephone can be installed for communicating with the CAD/CAM system operators.

Each workstation should have an associated desk where a user can place a model, or spread out a drawing, or make notes. It should be possible for a user to move a workstation nearer (or further away) and to rotate it upward, downward, or sideways to find the most convenient position. As a user may spend several hours a day at the workstation, it is important that seating be adjustable to the most comfortable position.

As the users will be working on key elements of the company's new products, access to the workstation room should be strictly limited.

4.3.2   CAD/CAM System Operations

Operational procedures must be put in place so that users can start
to use the system. Scheduling of initial system use, applications,
projects and users must be carried out. The standards to be used in
the system must be defined, entered in the system and made
known to the users. Similarly a parts library must be built up and
its contents made known to the users. The use of each specific
"layer" must be defined. Back-up and archiving procedures must
be implemented. A configuration manual should be developed.
It should include a description of how to use the system, the
standards and parts libraries implemented, rules for naming parts
and drawings and for describing the state of information (e.g.,
working sketch or released drawing).

4.3.3   Management of CAD/CAM Data

One of the key roles of the CAD/CAM team is to assist users and
their managers in increasing the usefulness of the system. One of
the most basic reasons for purchasing the system is to improve
information flow—both between different application areas and
within particular applications. The CAD/CAM system contains not
only a set of functions (to model a surface, to draw a fillet, to
create a tool path) but also a capability to handle data (i.e.,
information). The system's capability to handle information must
be exploited to the full. If a part has been designed, it should not
be redesigned from scratch—but how can a user know if it has
already been designed? Similarly, a part programmer wanting to
program a part that has been designed on the system needs to
know where to find it.

Data is managed in several different ways within the CAD/CAM
system:

Systems generally have log-on and password facilities to stop use
by unauthorized personnel. Passwords can be allocated to individual users and they can also be grouped so that one group of
users can create, access and modify its own data, and can access
but not modify the data of another group. Sometimes a mixed

system of passwords and priorities is used to enable selected
users to access data that would otherwise be inaccessible.

Many systems allow different types of data to be stored on different "levels." Thus a designer may enter geometry on level 1,
dimensions on level 2, extra texts on level 3. The part programmer may use level 4 for tool path geometry and level 5
for texts.

Most systems contain a library of predefined components expected to be used very frequently. Users may add often-used,
standard, company-specific components or parts to the library.
Afterwards, these will not need to be recreated each time that
they are needed, but simply recalled and placed in the correct
position.

Some systems offer possibilities to create families of similar parts
by modifying the geometric parameters of a predefined basic
family part geometry.

Most systems have a data management facility which, depending
on the system, stores a part, or a drawing of a part, or just a
drawing under a user-selected name.

Since systems have a limited on-line memory capacity, much data
will have to be archived off-line, often on magnetic tapes.

An average user may produce 200 geometric models and drawings per year. A company with four workstations could well have
10 such users and could therefore produce some 2000 models and
drawings per year, each of which could be linked to a file containing associated information such as texts, materials, quantities, etc.,
and could also refer to parts in a standard parts library. Unreleased
data, incomplete parts of models, and so on would increase data
storage requirements to well over 200 MB per year. Complications
arise when changes are made to existing data, in particular to parts
in a standard parts library. For new designs and assemblies, increasing use of parts of existing designs will occur. It is easy to see
that very quickly the amount of information and the complexity
of relationships between information in the system reaches a point
where the maintenance of adequate records by manual methods
becomes impossible. Many CAD/CAM systems do not currently

have good data management and retrieval facilities and it is left to the user company to develop its own data management system.

Apart from the CAD/CAM system users, the other people involved in data management are the user's managers, the data management system administrators, and the data management system operators. The responsibilities of the administrators include ensuring the availability and usability of the system, and the definition of the rules governing the use of the system. The operators ensure that the system is kept in good working order, help the users to use the system, and enforce the rules. The users' managers use the system for functions such as signing-off work and the production of reports.

The system must include facilities to store and catalog a finished model, drawing, or standard part, and to list the contents of the data base. Formal procedures must be included for changing existing models, drawings and parts, for handling version and change numbers supplied by the user, and for informing interested parties of changes. Procedures must exist to allow for the formal release of data, and to search for required information. As with manual techniques of data and information handling, it must be possible to define who has the right to create, protect, modify, and copy data. It must be possible to work down (or up) through the different levels of information, for example, from the name of assembly it should be possible to find all information (including detailed part drawings) associated with that assembly.

Part names and drawing names lie at the heart of the data management process. Part names can be selected as a function of the company's existing part numbering scheme. Companies with several part numbering schemes may use the introduction of CAD/CAM to standardize on one part-numbering system. After all, a major objective of CAD/CAM is the use of existing information to save time and money. Multiple part numbers have the opposite effect, as users often believe that it is quicker to redesign a part than to find it in an existing data base. Standardized parts classification can also be implemented as a step towards group technology.

4.3.4 CAD/CAM Education and Training

CAD/CAM knowledge, and efficiency in using CAD/CAM, should grow from the day that the company decides to purchase a system and should never stop growing, since the more a system is used, the more information goes into it and the more is known about how to use it. Education and training are necessary to ensure that growth is as rapid as possible, in particular in the early stages. CAD/CAM education offers a somewhat theoretical form of learning, its intention being to teach the general, fundamental principles of CAD/CAM, with perhaps a few associated practical illustrations of their application. CAD/CAM training offers a much more practical form of learning, its intention being that the user learns application-specific skills.

The basic questions that arise concerning CAD/CAM education and training are:

Who has to be taught?
What has to be taught?
How should it be taught?
When should it be taught?
How long does it take to learn?
Who are the teachers?
How much will it cost?
How does learning fit in with productive system use?
Where should training take place?

There are no standard answers to these questions, only company-specific answers. Before specific answers can be given, the company's particular objectives of CAD/CAM education and training must be defined. These will be related to the overall CAD/CAM objectives of the company, for instance, the company may want a particular application to be as productive as possible. In all companies it is useful to set up a means of testing the effectiveness of education and training in meeting the objectives.

Who Needs CAD/CAM Education and Training?

In general, three categories of staff need education and training—
the users, management, and the CAD/CAM team.

The users of the system need to know how to use it in their
everyday work. There will also be some superusers who need to
acquire a deep understanding of the system, and who the average
users will turn to for assistance with special problems that may
arise only once or twice a year. The users' managers (i.e., middle
management) will have problems managing projects and users if
they do not have any understanding of CAD/CAM. Middle
managers find themselves caught between users who know the
details of the system but do not understand the strategic implications of CAD/CAM, and top managers who know the strategic
implications and want to see results, but do not understand the
details of implementation. They will have to decide which users
need which type of training, which users will work on which
projects, which work will be carried out on the system initially
and which projects should be carried out by CAD/CAM techniques
and which by manual techniques. Data management will become
an important part of project management as users attempt to
access existing data in the system, and interact with each other via
the system and its database.

The second major category of staff which needs to learn about
CAD/CAM includes financial staff and top management. Financial
managers must be taught enough about CAD/CAM to ensure that
in early years it is not overcharged to the users, and thus stifled.
Top managers must understand not only the potential benefits of
CAD/CAM, and the problems that it will cause, but also how it
will affect apparently unconnected areas of the company (outside
of design and manufacturing) such as marketing. They must be
able to initiate and carry through the changes needed for CAD/
CAM to be used successfully.

A lot of education and training is required for the CAD/CAM
manager and the CAD/CAM team. They will be under a lot of
pressure, and when things go wrong, as they invariably do, the
CAD/CAM team is often the first scapegoat. It is essential that all

team members, including the CAD/CAM manager, learn how to use the system. The CAD/CAM manager who is asked to demonstrate the "new toy" to a potential client, and replies "I don't know how to" will not have much future in a dynamic company, especially if the client's order goes to another company.

What CAD/CAM Training and Education is Necessary?

In the previous section, it was seen that there are three major categories of people requiring education and training—the users, management, and the CAD/CAM team. Within each category, different people will require different education and training, for example a part programmer will not need exactly the same training as a designer, the marketing manager may not need the same education as the finance manager, and the CAD/CAM manager will need to learn different things to the CAD/CAM team members. To cope with all the different needs, a large number of courses needs to be offered, and individuals should follow the particular courses that are necessary for them to be able to carry out their functions effectively. Course topics include:

The benefits of CAD/CAM at the company level
A general introduction to CAD/CAM
A general introduction to the chosen system
CAD/CAM hardware and software
An overview of CAD/CAM use in mechanical engineering
A general introduction to using the CAD/CAM system
The state of the art in CAD/CAM
Use of the system for a particular application
Transferring data from one application to another
Operating the system
Developing new software for the system
Standard company practice for using the system
Data and drawing management procedures
Ensuring CAD/CAM is used effectively
An introduction to computers and graphics
Case studies in the use of CAD/CAM
Economic and financial aspects of CAD/CAM

Social aspects of CAD/CAM
Trade union attitudes to CAD/CAM
Procedures for selecting new projects for CAD/CAM
Procedures for selecting new users
Training new users
Developing the company's CAD/CAM data base
Management's role in developing CAD/CAM

It will be seen that there is a lot to be learned about CAD/CAM if it is to be successfully introduced into the company. The high requirement for training is typical of the introduction of automated technologies. New skills are needed and they must be learned. As automation and computerization increases, the need for training will also increase.

Within each of the three major categories, users, management, and CAD/CAM team, the question will arise of who to teach first. Some people will feel left out if they are not trained, others will feel victimized if they are trained. In the user category, the best solution is to pick the people who are expected to be the most efficient with the CAD/CAM system. It is necessary to explain to all concerned exactly what is happening, and why someone is picked for training at a particular time—presumably there is a reason and hiding it will not help progress to the ultimate goal of increasing efficiency. It is important however, that the first people selected for training have a desire to work on the CAD/CAM system, have good application skills, be enthusiastic and creative, and understand that the system is not a universal panacea. People with a strong dislike for computers, no interest in new ideas and concepts, and a negative approach to their work can be trained later. Among managers it is important to educate as soon as possible those who are going to have to make decisions concerning the system. As for the CAD/CAM team, all members must understand that they will be spending most of their time learning or training others, or in other completely new situations in which they will have little experience to help them. They therefore need as much training as possible as soon as possible.

How Does CAD/CAM Training Take Place?

The basic methods of CAD/CAM education and training concern large groups, small groups and individuals. Lectures and conventional classroom techniques can be applied to large groups. Seminars, workshops, and case studies apply to small groups. Training methods for individuals are often oriented to on-the-job training. The basic media of education and training are people, audio and visual tapes, books, films, manuals, and computers.

Only the very broad topics such as a general introduction to CAD/CAM and an overview of CAD/CAM use in mechanical engineering can be presented to large groups. These topics could be presented by consultants, or the system vendor, with the aid of advanced audiovisual means. The audience would include all managers and all users.

The majority of the topics will be presented to small groups, with individuals going on to further study either in the office (generally with written material) or on the job (generally at the workstation). The vast majority of a user's training will take place at the workstation, being guided by an instructor from the system vendor or the CAD/CAM team, or by a more experienced user. Many systems can be run in a training mode, in which they lead the user through an application showing, step-by-step, the different functions that are available.

The user's training usually starts with a set of seminars and demonstrations showing how the basic functions of the system work. The users are then given an overview of each command of the system and its effect. The purpose and use of the various graphics commands are explained. The user moves on to learn how to combine commands to produce a required effect. Rather than learning parrotwise, the user is generally expected to carry out exercises requiring thought as to which functions should be used. Often, some time is spent on learning how to operate the system and workstation, starting it up, requesting a drawing, etc. Some time will be spent on learning to design in two or three dimensions. The user will learn how to create a part and enter a part into a parts library. The user will learn how to model assemblies. By the end of the initial training period, it should be clear to the user that CAD/CAM is a tool that is as efficient as the usual manual tool, and has the potential for being much more efficient. If this is not the case, as often happens, then serious problems are not far away. Many users appreciate neither extended periods of formal education nor reading computer manuals. They prefer hands-on training, if possible from an instructor with a user background (most definitely not from a computer specialist).

The Timing of Education and Training

CAD/CAM education and training are ongoing processes that start as soon as the company decides to invest in CAD/CAM. At this early stage, the task force will learn fast. If possible, the CAD/ CAM manager should be designated before or when the task force is set up, and should be one of its members. Once a particular system has been selected, and before it is installed, substantial benefits can be generated by starting to train the CAD/CAM team and selected users. It is important both politically and technically that when the system is installed there are capable and proficient staff available to carry out tests and demonstrations. The people who have committed the company to CAD/CAM and a particular system will want to show as soon as possible that they have made the right choice. They will also want to attract as many users as possible while CAD/CAM has an attractive, new toy image. Carrying out initial training onsite is often inefficient because of system malfunctions in the early stages and other interruptions such as management wanting to see the system, or the trainees being called away for other, more urgent tasks.

Another reason for starting training as soon as possible relates to the delivery cycle of the system (usually between three and six months) and the long learning curve associated with becoming proficient with the system. Although many simple, essentially drafting, systems can be used productively after a few days, more complex systems may require six months or even one years use before full proficiency is attained. Rather than assuming that these two delays should be added together and thus cause an unproductive deadtime of about a year, selected users and the

CAD/CAM team should start training well before the system arrives. When the system finally arrives, these users will be able to train other users, they will be available immediately for productive work, and the work that they have carried out during their training will serve as a basis for use of the system. Training in advance of system installation can be carried out by renting a system, by installing a secondhand system, by connecting to a service bureau, or by sending users to another site.

After the system has been installed, education and training should continue on an expanding basis. Not only does existing knowledge have to be disseminated to other people, but further training is necessary to develop new techniques.

The Duration of CAD/CAM Training and Education

The company must see CAD/CAM training and education as a continuous, neverending investment. New users will have to receive basic training, experienced users will need advanced training, and in managerial roles, staff must always be learning more so as to be able to increase the benefits derived from CAD/CAM.

Training periods for users depend very much on the complexity of the system. Whereas some systems may require only one or two days of initial training, others may require two or three weeks. Some companies, wishing to ensure that the users really master the principles of CAD/CAM and the system, give up to two months initial training. In other cases, training in the first year is split into two periods, with perhaps 70% given initially and the other 30% being given when the user has reached a certain level of competence. Depending on the system and the application, a user will probably require about three months training and use of the system to become as efficient with the system as without it. In other cases, this period in which the user is less productive with the system than without it, may extend to six months. To reach maximum efficiency will, depending on the system, require between six months and one year. These times may appear to be long, but are realistic. Once accepted, a company can try to reduce them by improving education and training. The alternative, an ostrichlike refusal to see the need for education and training, will soon lead to nonoptimal system use.

Who Gives CAD/CAM Training and Education?

There are basically five sources of training and education:

1. Consultants
2. The system vendor
3. Internal (e.g., the CAD/CAM team)
4. Educational institutes
5. CAD/CAM training companies

One of the major roles of consultants is in educating management about CAD/CAM. The consultant's overall, neutral, general knowledge of CAD/CAM can also be used at the system selection stage, and in aiding the company to decide which applications should be transferred to CAD/CAM.

The system vendor will offer the CAD/CAM team basic training in how the system works, and will offer basic and advanced training courses for the users. The cost of the basic courses is often included in the overall price of the system.

Internal sources of training and education (e.g., from the CAD/CAM team and from superusers) will become more and more important as the amount of internal knowledge increases and as the overall system takes on a company "flavor." The company flavor results from the fact that since each company has different requirements, users, applications, and skills it will use its system differently from other companies.

Many educational institutes offer CAD/CAM courses. These represent a good way of gaining a general CAD/CAM education. Such courses can be especially useful during the stage in which the company is trying to raise the general level of CAD/CAM knowledge.

CAD/CAM training companies (and institutes) offer courses for users of specific CAD/CAM systems. An instructor who has been with such a company for some time and who has trained users from many different industries and application areas, can give extremely efficient training in both basic and advanced courses. Long-term contacts between such an instructor and a user company are particularly productive as they give the instructor the opportunity to get to know the company requirements in depth.

Costs of CAD/CAM Education and Training

The cost of education and training depends on variables such as the ability of the people involved, the particular system, and the degree to which the system is interfaced to other systems in the company. It should not be forgotten that in-house training ties up otherwise productive workstations. The overall annual cost of education and training is generally at least 10% of system (hardware and software) cost.

Matching Training and Productive System Use

Wherever possible, training should be given in such a way that it has a visibly productive result. This is important both from the financial point of view, and to provide motivation to the user.

During initial training, before or at system installation time, users can input existing parts that will form part of the company data base. Alternatively, they can work on low-priority projects where time is not critical, and major blunders will not be too expensive.

Once the system has been in successful operation for six months or a year, there should be no shortage of test examples for new users to cut their teeth on, or of productive work that can be overseen by an experienced user.

4.4 CAD/CAM SYSTEM INSTALLATION

With good planning, the different sites will have been prepared by the time that the vendor is ready to deliver the system. As soon as the equipment arrives, the vendor's staff should unpack it and install it. Hardware and software acceptance tests should be carried out together by the vendor's staff and the CAD/CAM team. If the results of the tests are acceptable, then work on the system can start. If results are completely unacceptable, the CAD/ CAM manager should inform top management of this, and also inform the management of the system vendor by registered letter. However, in general, the results of the acceptance tests will be neither in complete agreement nor complete disagreement with what was promised, and it will be up to the CAD/CAM manager to persuade the vendor to bring the system up to the expected level as soon as possible.

4.5 INITIAL USE OF THE SYSTEM

At this stage, if all has gone according to plan, the company will at last have something to show for its investment in selecting and planning the implementation of CAD/CAM. Initial system use may show that some of the plans do not work as well as expected in practice, and corresponding refinements may be needed. Use of the system should be monitored from the very beginning.

5

ONGOING CAD/CAM OPERATIONS

Many of the activities (including training, selecting and installing new hardware, building up the data base) that take place during system selection time, prior to installation, and at installation time, continue throughout the lifetime of the system. However, other new activities only really appear after the system has been installed for some time. They can be grouped into the following major areas:

Systems operations
Organizational issues related to system use
Forward planning

Table 5.1   Activities in Ongoing CAD/CAM

System Operations
   System maintenance
   Flexitime or shift work
   New versions
   Back-up
   Workstation scheduling

Organizational Issues Related to System Use
   User meetings
   CAD/CAM system documentation
   Company-specific CAD/CAM manual
   User groups
   System monitoring
   Implementation of standards
   Data
   Project management

Forward Planning
   Integration with other systems
   Definition of a career path for CAD/CAM specialists
   System expansion
   Learning from others

Managing the Implementation of CAD/CAM
   Middle management involvement
   Management of change

Auditing the Implementation of CAD/CAM

Managing the implementation of CAD/CAM
Auditing the implementation of CAD/CAM, as shown in Table 5.1.

5.1   SYSTEM OPERATIONS

5.1.1   System Maintenance

Maintenance of the CAD/CAM system should be arranged to take place when it will cause the least possible disturbance—if possible at the weekend or at night, otherwise early in the morning or late in the evening. A CAD/CAM system is an expensive resource and should be used productively as much as possible.

5.1.2 Flexitime or Shift Work

With standard company working hours of eight in the morning to noon, and one o'clock in the afternoon to five o'clock, expensive workstations can only be used for a maximum eight hours per day. To avoid such a waste of a valuable resource, the solution will almost certainly be a multishift, flexitime approach, with workstations being available and used from six o'clock in the morning to ten o'clock in the evening (i.e., 16 hours per day). In the future, another solution may be to run applications with limited requirements on low-cost workstations and personal computers attached to the system. While investigating the multishift approach some thought should also be given to allowing users to work at the weekend if they want to. System response time is good at the weekend and distractions from other staff are minimal.

5.1.3 New Versions

Every four to six months, a new "version" of the system software will be delivered by the vendor. Users will have to be trained to handle new and modified functions. To avoid wasting too much time on such training, some companies only implement the new versions of the system once a year. Users will therefore only have to be retrained once a year. Unfortunately, new releases are sometimes not compatible with existing data files and interfaces to other programs. This causes a lot of extra work for the CAD/CAM team. Vendors should be requested to maintain compatibility with existing versions.

5.1.4 Backup

In case of severe system hardware or software failure, all data (parts, libraries, programs, etc.) currently on the system can be lost. It is therefore usual to copy regularly information held on the system to a back-up medium—generally a magnetic tape. In case of system failure, the back-up tape can be read back into the system and work can continue.

5.1.5 Workstation Scheduling

Although every user of the system would like to use a workstation whenever necessary, this is often not possible, because it could mean that when the user did not need the workstation, it would be unused and therefore losing money. Most companies introduce workstation scheduling so that the users can plan their sessions at the screen (of one, two, three, or four hours) in advance. Scheduling also assists in ensuring that users are forced away from the screen from time to time. It is only too easy for a user to try to resolve problems at an expensive workstation that would be better (and more cheaply) resolved back at the desk or at a low-cost terminal.

5.2 ORGANIZATIONAL ISSUES RELATED TO SYSTEM USE

5.2.1 User Meetings

The users of the system should be encouraged to hold two or three meetings a month. The aim of the meetings is to make the users more efficient in using the system, since if they are inefficient, the system however expensive and powerful it may be, will not produce the expected results. One meeting could concentrate on special, time-saving techniques that have been discovered, another (in the absence of management) could deal with complaints. A third meeting, with both users and management present, could aim to develop ways of improving overall system throughput and efficiency.

5.2.2 CAD/CAM System Documentation

As users become more and more experienced they will need to know more about the way in which the system works. One way for them to learn is by consulting the system documentation. Often this documentation makes no sense during early stages of system use, but becomes more valuable as the user gains experience. Enough documentation should be available for users to be able to consult it when required.

5.2.3 Company-Specific CAD/CAM Manual

The CAD/CAM system documentation explains how the system works as a system, but not as a part of a particular company. The company must produce its own documentation defining standards, standard parts, operating procedures, naming conventions, etc. Details must also be available describing how to use the CAD/CAM system in conjunction with other systems in the company.

5.2.4 User Groups

User groups, with members from many companies using the same CAD/CAM system, exist for most systems. As well as giving users and the CAD/CAM team a well-deserved visit to California or the South of France, user group meetings are useful forums for learning how others handle the same system, and for uniting with other users to pressure the vendor into correcting bugs and carrying out much-needed modifications and new developments.

5.2.5 System Monitoring

Most CAD/CAM systems have some automatic features for monitoring their use and producing corresponding documents. The CAD/CAM manager must develop ways of monitoring use of the system in relation to company objectives. It is essential to know if the system is providing the expected benefits. If it is not, then it is necessary to find out the reason, and to take corrective action.

5.2.6 Implementation of Standards

Nearly all companies make some use of standards, whether their own, national or international. Exactly the same standards must apply to work carried out on the CAD/CAM system as to that carried out manually. Standards offer a major source of savings and some companies use the introduction of a CAD/CAM system as an opportunity to tighten up on the use of standards. It will take time to produce a standards data base, and the most effective standards should be established first. Once established, they must be implemented, documented and made readily available for use.

5.2.7 Data

As use of CAD/CAM spreads throughout the company, problems concerning data management, communications, privacy, and security will increase. Data will be reused at different stages of the design and manufacturing cycle. This may require transferring data between different CAD/CAM systems, between different departments, and perhaps between different companies. To carry out such transfer efficiently requires a standardized CAD/CAM data exchange format. The standard formats proposed so far (e.g., IGES) cannot efficiently transfer all types of geometry, and do not completely define other information concerning an object such as topology, attributes, and relations. In the long term, standardized access to common CAD/CAM data bases and commercially available standard parts libraries (equivalent to today's catalogs of parts) will be required. A CAD/CAM system will then be able to access any one of several different parts libraries. Similarly as new CAD/CAM systems and new versions of existing CAD/CAM systems are produced, it is important to check that it is still possible to recall and handle archived data.

5.2.8 Project Management

Engineering projects have always required good project management. The use of CAD/CAM on an engineering project will increase the need for good project management since new resources are involved, new procedures must be applied and the overall time cycle will be reduced. It is very important during early phases of CAD/CAM use (when both users and managers are unfamiliar with the system and the procedures) to keep project progress under close scrutiny. It should not be forgotten that much project management, prior to installation of CAD/CAM, works because people are used to it and have been doing it for years. They will have to be trained to have the same reflexes with CAD/CAM.

5.3 FORWARD PLANNING

5.3.1 Integration with Other Systems

Initially a company will probably only use CAD/CAM on a restricted number of applications. Once these are running successfully, other applications must be switched from manual methods to CAD/CAM. Gains will occur if these applications make use of data already in the CAD/CAM data base. There is an ongoing requirement to identify applications to be switched to CAD/CAM, and to ensure their successful transfer to the CAD/CAM environment.

5.3.2 System Expansion

It is virtually impossible for a company to switch all its applications to CAD/CAM at the same time, or to train all its users simultaneously. Implementation of CAD/CAM will therefore take place in stages. As each new stage is reached, new applications will be switched to CAD/CAM, new users will start to use the system, and other systems will be interfaced to the CAD/CAM system. New hardware and perhaps new software will have to be installed.

5.3.3 Learning from Others

The implementation of CAD/CAM is by no means terminated when the system has been installed and is being used productively. The use of CAD/CAM within the company will continue to grow and grow. A good level of CAD/CAM knowledge must be maintained within the company, or the use of CAD/CAM will rapidly become nonoptimal. One way to keep knowledge levels up is to go to user group meetings and CAD/CAM conferences. Another way is to visit CAD/CAM users and managers in other companies.

5.3.4 Definition of a Career Path for
CAD/CAM Specialists

There is currently a worldwide shortage of CAD/CAM specialists, and a CAD/CAM team member who works hard and well will soon receive interesting offers from vendors and other user companies. To hold on to such staff it is necessary to show them that they have a well-remunerated, interesting and important future with the company.

5.4 MANAGING THE IMPLEMENTATION
OF CAD/CAM

5.4.1 Middle Management Involvement

Middle managers who must make sure that projects meet well-defined objectives in the optimal way can make or break a CAD/CAM system. If they do not want to get involved in CAD/CAM then, no matter how involved top management and the users are, maximum productivity will not be obtained. It is therefore very important to involve middle management in CAD/CAM and make their responsibilities highly visible and very well defined.

5.4.2 The Management of Change

As CAD/CAM is progressively installed in a company, it will lead to changes in the way that work is carried out and in the responsibilities associated with this work. Top management must ensure that changes take place so that CAD/CAM, and its associated store of information, can be used as effectively as possible. On a purely technical level, these changes can be implemented by the CAD/CAM manager. However, as soon as they are no longer purely technical—when they involve the organizational structure and staff, it is only top management that can ensure their implementation. It is obviously preferable for management to be aware of the need for changes and to set them in motion slowly but surely, than to wait until they cannot be implemented without dramatic consequences on staff, on staff morale, and on ongoing work.

Management cannot expect changes to occur without definite actions being carried out. People get used to "the way things are," and will try to keep them that way. After all, change may result in loss of power, a requirement to learn new techniques, and the need to work harder. Although top management may introduce CAD/CAM with the sole intention of improving productivity, many people throughout the company will see it as a personal attack on their current way of life. Management must act positively to overcome such attitudes.

5.5 AUDITING THE IMPLEMENTATION OF CAD/CAM

Now and again it is useful for the company (or perhaps a consultant) to evaluate the real status of CAD/CAM in the company. The original CAD/CAM plan may have been excellent and everyone may have tried to adhere to it, but several years under everyday pressure to produce the company's products may well have led to some differences between where the company should be in CAD/CAM, where it thinks it is, and where it actually is. A quick but formal check may well unearth some potential for rapidly increasing productivity.

6

THE FUTURE OF CAD/CAM

I've been asked to make a little speech to you about the way CAD/CAM has evolved. You don't need to worry too much about the individual technical details of what I'm going to say. It's the overall picture that's more important. Most people in CAD/CAM spend too much time on short-term problems. My speech should stretch your horizons a little, and in doing so should help you to change your CAD/CAM focus.

Let me introduce myself. I'm Yves Grotzer. It's now 20 years since my father Wayne Grotzer bought the first CAD/CAM system for Golden Triangle Furniture. The company has changed a lot since then. Once those pioneers had seen what CAD/CAM could do with wood furniture, they diversified into plastic furniture and then into all sorts of plastics engineering. The company has grown an average 15% per year over the last 20 years, which means we are now about 15 times as big as we were back in 1985. We still have the main offices and plant here in Pittsburgh, but we also have plants in Florida, Texas, California, and the state of Washington. Anyway, I'm digressing from the subject. I was asked to tell you about the way CAD/CAM evolved from 1985 onward. Well, I would say there was vertical evolution, horizontal evolution, and superficial evolution. Vertical evolution is about the way an individual function, system, or piece of equipment is improved, and I'll be saying more about that later. I'll also tell you more about horizontal evolution—the way systems that appear to be separate, change their functions and interfaces, and eventually finish up as one system.

6.1 SUPERFICIAL EVOLUTION

Let's start with superficial evolution. First of all the name of CAD/CAM changed. I don't think anyone ever really liked the CAM in CAD/CAM since the subject is not so much about manufacturing in general, but about manufacturing engineering. Someone suggested that CAD/CAM should be called CADE/CAME for computer-aided design engineering and computer-aided manufacturing engineering. Then it became CADCAM. Eventually it settled down as CAE, computer-aided engineering and it was very clear that both design engineering and manufacturing engineering were being referred to.

In the late 1980s, quite a change occurred in that a lot of manufacturing engineering and mechanical engineering companies really started to use CAD/CAM. I guess that before then it had been difficult to cost justify, but as costs came down and functionality increased, the day came when a lot of companies could really justify CAD/CAM system purchase. Before then, there had been a lot of talk from the vendors, but only a few brave companies had got their feet wet. The fact that these companies really started improving productivity through CAD/CAM also helped to get the ball rolling.

A lot of the early system vendors went bankrupt or were taken over, and now more than 90% of the market is in the hands of three big vendors. There are always new companies starting up, but they tend to concentrate on giving very good service in one geographical or engineering area, or they bring out advanced systems with new technologies before the big vendors can get their act together. Invariably, if the new company is any good it gets swallowed up by a larger company.

Before I move on to vertical evolution, let me remind you that we have a CAD/CAM Industrial Museum here in Pittsburgh. My father first thought of building it up after he had visited an automobile museum near Detroit. Let me tell you the story. When my father was about to buy our first CAD/CAM system, a lot of his friends tried to dissuade him. Some just wanted things to carry on being done in the good old traditional way, others were worried about the data base exploding, one thought the users would go blind, another said that systems were too difficult to use. After my father had visited that car museum, he saw the similarity between the early days of motoring and of CAD/CAM. The early users of automobiles knew that they would have a few problems to start with but they also knew, although they could not always justify it financially, that in most cases, an automobile would be able to outperform a horse.

I remember how my father used to tell me that there was a long learning period, of between two and four years, during which a company got to know how to handle CAD/CAM effectively. A company that did not get into CAD/CAM was not gaining anything, just putting back the date at which it would use CAD/CAM effectively. My father used to compare this to the horseman who preferred to wait for the next car model (you know, the one that really works). Even when the new model came out, the horseman still didn't know how to handle it, whereas by that time the people who had bought the first model were using it very productively.

When he set up the museum, my father did not just want people to come and gape and laugh about the way those old fools had worked, or to reminisce about the "good old days." No, he felt that there were far too many people in industry who took decisions on technologies they did not understand. He wanted those people to see and realize just how quickly a technology can develop, change, and grow to have a strategic importance.

6.2 VERTICAL EVOLUTION

Way back in the 1980s, when CAD/CAM was still in its infancy, people concentrated too much on details—they would say "Look, now you can draw in color on my workstation" or they would categorize computers as microcomputers, mainframes, superminicomputers, and pretend there was some kind of difference. The only real difference was in the computing power and as that was growing at an enormous rate, one year's superminicomputer could be more powerful than the previous year's mainframe. Of course, things changed when fifth-generation computers arrived, but that's another story. The real details seemed to be forgotten about, for example, the number of colors available was very limited and often different colors could not be handled by the CAD/CAM software. The screen size was often very limited, and the workstation was not intelligent, took up an excessive amount of space, and could not be adjusted to the best position for the user.

When you look at today's workstations, or workplaces, as we now call them, you begin to feel that they were built for adult engineers not for ten-year-old kids. To start with, the screen looks something like an old-fashioned drawing board. It is flat, measures about six feet by three feet, can be adjusted to the best position, and can display as much information at a time as the engineer requires. Each workstation has its own built-in computer which is about ten thousand times as powerful as the "superminis" of the 1980s. The workstation computer is connected via a network to the other workstations and to the major computing resources such as the data base machine. All data is generally held online, and most companies use automatic backup onto disks. The CAD/ CAM systems have really powerful data bases now. One of the problems in the 1980s was that you could put a lot of information into the system, but it was very difficult to do anything with it once it was in. Generally the number-crunching machine in the network is about ten million times as powerful as a 1980s "supermini." A lot of NC machines, robots, and automated handling systems are now capable of carrying out very intelligent work and have correspondingly powerful controllers. New systems have been developed so that this equipment can be programmed efficiently.

Artificial intelligence (AI) techniques progressed a lot slower than expected, but since 1995 they have been widely used. They are used all the way through the design engineering and manufacturing process. Given the specifications of a new product, AI techniques can be used to check that it meets specifications. From a knowledge base of process plans, tools, and fixtures for existing products, AI can be used to select appropriate plans and tools for the new product. Let me give you some specific examples of the use of AI. In roll forming, it selects the best set of die shapes. In plastic moulding, it selects the optimal position of runners and gates. In machining, knowledge bases of NC paths, tools, materials, feed rates, speeds, depths of cut, and so on are used automatically to select the best machining parameters for a part. In robotics, complex problems in programming robot kinematics for material handling and assembly are resolved with AI. In quality control, knowledge bases are used to define which are the most suitable product parameters to be checked. Artificial (or machine) vision systems are programmed from knowledge bases linked to CAD/ CAM data bases for applications such as location, identification, and routing of parts.

In the 1990s, advanced engineering analysis methods were developed using the more powerful computers and techniques available. In particular, geometric modelling software was greatly improved. We use solid modellers for all types of parts. Most of the basic CAD/CAM functions are now available in hardware. There's a company somewhere in Colorado that sells a full drafting system on a chip to one-man shops. With the screen and documentation, the system sells for less than $200. Another thing that happened in the 1990s was that the use of personal computers and low-cost terminals increased to the point where nearly everyone in the company had their own computer or terminal. A lot of training had to be given to ensure that people used systems productively. As systems keep evolving, the requirement for training does not go away.

Perhaps one of the biggest differences between now and the 1980s is the way that the user communicates with the system. The old light-pen and those unreadable menus no longer exist. The old alphanumeric keyboard is not used much as most people find it is much easier just to say what a data value is, rather than to type it in one character at a time. The user can talk to the system or give commands by touch, and the system replies "verbally"—asking questions, making helpful comments, or just confirming that it has understood what the user said or did. The system is equipped with "eyes" (or more correctly, a vision system) and sensors which follow and understand movements of the user's hands. The user has the possibility to identify a part of the model on the screen by pointing, or indicate how the model on the screen should be modified, or just use his hands to show the system what the part should look like. Maybe I should not have said "his" there, since the user could well be a "she." There are a lot of women in engineering these days. By the way, you may remember Anna Bellinaso who was on my father's task force when he first wanted to select a CAD/CAM system. She is now our Information Director.

That brings me to another point. When computers were first introduced, many companies appointed an EDP manager or an EDP director. To be honest, many of those people were too closely linked to finance or to the computer hardware to be really effective in an engineering and production environment. In fact, when CAD/CAM was introduced, a productivity director was sometimes appointed—particularly if the EDP director knew nothing about manufacturing. After some time, the amount and value of data produced by a CAD/CAM system grew so much that a data director had to be appointed to make sure that everyone got the right data in the most efficient way. The latest trend is to appoint an information director—after all, what CAD/CAM really does is to try to handle information in the most efficient way possible. The information just builds up, year after year, and many companies now regard it as one of their most important resources—as important as capital, staff, and equipment.

Only a limited amount of work is carried out by designers working from home. A few years back everyone was doing it, but some engineers working from rented accommodation in the Bronx managed to beat the security codes, passwords, and so on, and steal all the information on the country's latest space vehicle. That really opened people's eyes, and now access to information bases is very closely controlled. Of course, engineers can still do some work at home, but they can no longer connect into the company network.

6.3 HORIZONTAL EVOLUTION

Looking back on CAD/CAM in the 1980s, one can see that it was just a step on the way to the complete computerization of the manufacturing process. This process starts when a customer (or marketing) requests a particular product, and it goes through to delivery of the product to the customer, and to after-sales service. Computers were first used quite independently of each other on the shop floor, in manufacturing planning and control, and in engineering. Even within engineering, computers were first used quite independently of each other in applications such as part programming, finite-element analysis, and drafting. CAD/CAM in 1985 can now be seen as a grouping together of computer-based design engineering and manufacturing engineering functions, with the common link between these functions being the part geometry.

CAD/CAM expanded out of its traditional areas when vendors added modules to carry out computer-aided process planning, to automatically produce bills of material, and to extend part programming functions to handle DNC. To make these applications as efficient as possible there was growing use of group technology with its associated coding and classification techniques.

While CAD/CAM was spreading into other areas, the computerization of these other areas was also increasing. Over in the manufacturing planning and control area, the MRP specialists thought that the bill of material was "theirs." To increase sales, MRP vendors started to include CAD/CAM functions in their systems. Down on the shop floor, suppliers of machine tools added CAD/CAM functions to their systems, and shop floor control system vendors added functions that had previously been thought of as belonging to MRP to their systems.

Numerous problems arose for companies using these systems if a clear information handling strategy had not been established. Each individual system seemed to mushroom up and outward, interfering rather than interfacing with all adjacent systems. If care was not taken, different systems would produce competing versions of the same information.

Eventually of course, standards were defined for transferring data between systems, not only in CAD/CAM, but throughout the computerized manufacturing company. The companies with a clear strategy then had several solutions. They could use a single supplier solution, and for many small- to medium-sized companies this was the most attractive solution. The medium- to large-sized companies took the modules of the available systems that they required, and put them together in the way that best suited their mechanical engineering and manufacturing engineering requirements.

6.4 CLOSING REMARKS

Well, I hope my little speech has given you an idea of the way CAD/CAM has developed. There are two more things I think I should say. I half-remember a quotation from the French philosopher, Montaigne, I think. He said something to the effect that a ship that did not have a port of destination did not have much chance of making a successful voyage. It's pretty much the same with CAD/CAM from what I've seen. If a company's management does not know where its going with CAD/CAM, then that company does not have much chance of succeeding. Looking back, even though CAD/CAM technology has changed a lot, the needs for management commitment to CAD/CAM, planning, target setting, and the monitoring of results have remained constant.

The other thing I wanted to say was about people. Two people who really helped us to get into CAD/CAM were Eddie Ludd and Paul Kim. Before joining us they had been doing some CAD/CAM work for a company down in McKeesport, but things did not seem to work out down there. Both of them are still with us. Paul Kim is the best user we have ever had, and Eddie now manages CAD/CAM operations throughout the company. Once I asked them how it was that CAD/CAM had worked so well for us and yet so badly down in McKeesport. They told me that technology alone does not ensure success. They say that to be successful you have to have the right technology, the right people, and the right management approach. I'm sure that they are right.

Part II

Overview of CAD/CAM
Applications in Industry

7

OVERVIEW OF THE INDUSTRIAL USE OF CAD/CAM

In the first part of this chapter a brief description is given of the way in which CAD/CAM has been introduced into industry. Then, the current use of CAD/CAM in mechanical engineering and manufacturing engineering is outlined. Its application to the aerospace, automotive, offshore, ship, machine tool, footwear, power generation, clothing, hand tool, household appliance, packaging, machinery, and electromechanical component industries is summarized. Details are given of how CAD/CAM is used in forging, plastic moulding, NC machining, and sheetmetal working. Finally, a brief account of some national and international efforts

Table 7.1 Some CAD/CAM-Assisted Manufacturing Technologies in Industry

Manufacturing technology	Manufacturing tools	Typical industries using CAD/CAM for this technology
Metal forming	Forging die	Hand tool
	Casting mould	Automotive
	Stamping die	Automotive
	NC pipe bender	Aerospace, offshore
Plastic forming	Injection mould	Domestic appliance
Glass forming	Mould	Automotive, container
Rubber forming	Mould	Automotive
Leather forming	Shoe last	Footwear
Composite forming	Mould	Aerospace
Metal cutting	NC flame cutter	Ship, offshore
	NC mill	Power generation
	NC lathe	Machine-tool
	EDM machine	Electrode-making
	NC router	Aerospace
	NC pipe cutter	Offshore, ship
Leather cutting	Water-jet cutter	Footwear
Fabric cutting	Laser cutter	Clothing
	Knife	Footwear, clothing
Composite cutting	Water-jet cutter	Aerospace
Paint spraying	Robot	Automotive
Composite laying	Tape layer	Aerospace
Assembly	Arc welding	Automotive
	Riveter	Aerospace
	Sewing machine	Footwear

to promote the use of CAD/CAM is presented. Table 7.1 shows some CAD/CAM-assisted manufacturing technologies and typical industry sectors in which they are used.

7.1 A BRIEF HISTORY OF THE USE OF CAD/CAM

Computers have been used in design engineering and manufacturing engineering since the early 1950s. However, early uses, such as numerical control programming and finite-element analysis were batch processes and cannot be considered as CAD/CAM. It has been seen that the distinguishing features of CAD/CAM are the use of interactive graphics, geometry modelling, and the reuse of product information stored on the computer. Before interactive graphics could be used, the graphics terminal had to be invented and manufactured. Similarly, before product data could be reused, ways to represent product geometry had to be discovered and implemented.

7.1.1 1960–1965

During the period 1960–1965, these two activities were under development. At that time, the only computers commonly available were mainframes (in particular IBM and CDC), and the graphics terminals under development used the vector refresh technology. The expense of the hardware, the cost of software development, and the limited availability of the new hardware technologies restricted the number of CAD/CAM users in this period to the major U.S. aerospace and automotive manufacturers.

7.1.2 1965–1970

From 1965 to 1970, the fundamentals did not change. An expensive mainframe computer and an expensive vector refresh terminal were still the only equipment available for CAD/CAM use. There was still no CAD/CAM software available for purchase, so CAD/CAM users had to develop their own software. During this period, the use of CAD/CAM spread slowly through the U.S. and European aerospace and automotive industries, the Japanese automotive industry and the shipbuilding industry. In 1970, the use of CAD/CAM in mechanical engineering and manufacturing industries was limited to about 50 companies worldwide. Within these companies, CAD/CAM was only used on specific applications (e.g., the design and NC machining of complex surface parts). The percentage of parts designed and manufactured with CAD/CAM in these companies was very low, on average less than 1%.

7.1.3 1970-1975

The period of 1970-1975 was marked by the use of two newly developed technologies, the minicomputer and the storage graphics screen. The minicomputer was much cheaper than the mainframe computer and the storage screen was much cheaper than the vector refresh screen. Software that was not specific to one particular application was developed, and turnkey systems (made up of a minicomputer, a storage screen, and basic software) were marketed. The software was mainly for drafting, but sometimes had a limited 3D wireframe capability. This period saw the growth of companies like Computervision and Applicon. The use of CAD/CAM was still restricted to a small number of companies in the mechanical engineering and manufacturing engineering sector. To the average company, the cost of mainframe-based CAD/CAM systems was extremely high, while the cost of a turnkey system was very high. The majority of turnkey systems went to companies in the aerospace and automotive sectors.

7.1.4 1975-1980

It was only in the period 1975-1980 that the use of CAD/CAM really began to spread outside its traditional boundaries. The period was not marked by any great technological breakthroughs so the reasons must be looked for elsewhere. The improvement and improved reliability of turnkey systems and increased marketing efforts by the vendors of these systems, are probably important factors. The performance of computers increased while their price decreased. The many conferences and publications on CAD/CAM (all of which described benefits and few of which described the problems) gave it some sort of respectability. Without fully understanding all the implications, management began to see CAD/CAM as an answer to increased productivity. It was a period in which an "engineer" could justify purchase of a small installation (e.g., a two or three screen turnkey system) on the grounds of figures that gave the typical "manager" no real basis for decision. (At that time, how could a manager with no prior experience of CAD/CAM know if a CAD/CAM system was going to increase productivity by a factor of three to one, five to one, or seven to one, especially if the system salesman was saying it would increase by a factor of ten to one.) By the end of the period 1975–1980, the automotive and aerospace manufacturers all had at least one CAD/CAM system and were busy looking for other CAD/CAM systems for particular application areas. The majority of the power generation equipment manufacturers, the major aerospace and automotive suppliers and major ship designers and builders were also using CAD/CAM techniques.

7.1.5   1980–1985

The period 1980–1985 was really the first in which CAD/CAM was used in a wide range of mechanical engineering and manufacturing engineering applications. Several reasons for this can be identified, including changing technologies, a new approach to CAD/CAM software, the growing weight of experience, the growing need for productivity improvements, pressure on suppliers, increased competition, and so on. Whereas in 1970–1975 new possibilities were offered by the introduction of minicomputers and storage tubes, in 1980–1985 new possibilities arose from the introduction of microcomputers and raster screens. Although it may be said that both of these really just represented a continuation of the trend to cheaper, more powerful electronics, to the end user they were seen as a quantum jump. CAD/CAM systems were developed for use with black and white (or color) raster screens, and mainframes, minicomputers, and microcomputers. Microcomputers and raster screens were packaged to produce the single central processing unit (CPU), single-screen "engineering workstation." The engineering workstation had a better price-performance ratio than a single-user turnkey system or a single-user mainframe system. It was therefore of interest both to small companies with a small number of users and, as a result of net-working capabilities, to large companies with many users who had problems with overloaded turnkey and mainframe-based systems. The introduction of engineering workstations and personal computers led to the development of new software to run on these computers. This software was built using advanced techniques and with the hindsight gained from the development of earlier systems.

The price of this software tended to match the price of the hardware, and was therefore much cheaper than the turnkey system software. Other factors affecting software during this period were the introduction to the market of several systems developed in Europe that challenged what had previously been a U.S. supplier-dominated market oriented to drafting. The European systems tended to be more oriented toward geometric modelling. It was in the period 1980–1985 that solid modellers began to be used, albeit rarely, for everyday industrial applications.

The increased need for productivity improvements, the lead given by governments for investment in CAD/CAM, and the widening use of CAD/CAM all led management to the conclusion that CAD/CAM really was necessary. In many cases, it was realized that there was no alternative to investing in CAD/CAM. When a car manufacturer no longer transferred a drawing to a supplier, but a tape containing data from a CAD/CAM system, the supplier was virtually forced to use CAD/CAM techniques. Similarly, the company that found that its competitor was using CAD/CAM techniques to bring products to market in 6 months rather than in 18 months had little choice but to invest in CAD/CAM.

The period 1980–1985 saw the use of CAD/CAM move out of its traditional "high-tech, big spender" home in aerospace and automobile companies and into companies involved in everyday, "low-tech, low-spender" mechanical engineering. At the same time CAD/CAM began to be used in virtually all manufacturing technologies.

7.2 CURRENT USE OF CAD/CAM IN VARIOUS INDUSTRIAL SECTORS

7.2.1 The Aerospace Industry

In the early 1960s, the aircraft industry was one of the first users of CAD/CAM. By 1985, CAD/CAM was a regularly used tool throughout the aerospace industry. Aeroplane, helicopter, rocket, and satellite manufacturers all take advantage of it, as do suppliers of parts as diverse as engines, tires, and seats. CAD/CAM is also used in cabin layout.

CAD/CAM techniques are used at many stages of the design and manufacturing engineering cycles of an aircraft. As early as the conceptual design phase, fuselage and wing geometry can be defined and held in the system's data base. This geometry is then input to aerodynamic and structural analysis programs and eventually becomes the basis of NC programs to produce wind tunnel models.

In more detailed stages of design, CAD/CAM techniques are employed in defining the exact shape of the aircraft, in positioning antenna and radomes, in space allocation studies, and in producing structural drawings. As interference checks can be made with the system, it is possible to build to a tighter tolerance, and still have fewer rejections. Improved geometry and documentation are of benefit to subcontractors trying to understand a design. Parts lists are input through the system keyboard. The system is also used to assist in defining the position and connections of the engine and the associated power system, the fuel system and the environment control system. CAD/CAM techniques are used in carrying out kinematic analysis of moving parts such as wing flaps, landing gear, and door mechanisms. The position and logic of the miles of wiring that go into modern aircraft is decided with the assistance of the system. Three-dimensional anthropomorphic models are built up in the computer and used in testing the freedom of movement of crew members. Clearance and vision can also be checked. Cockpit instrument panels are laid out with CAD/CAM. Tubing is designed taking care that sufficient clearance exists.

In the manufacturing engineering area, process planning may be assisted by a computer-based system that is either part of the system or closely linked to it. Users of the system generate part programs for machining centers and other NC-controlled machine tools. Structural parts are developed (i.e., flattened out) by the system so that the flat shapes can be cut out of large sheets by NC-controlled sheet-metal cutting machines. NC tube-bending machines form tubing to the required dimensions. Tools, jigs, and machine tool fixtures are designed with CAD/CAM. Multilayer composite parts and their forming tools can be designed with the system. NC-controlled quality control machines are programmed directly with CAD/CAM geometry, as are NC-controlled tape-laying machines, robots for applications such as drilling, and electrical testing machines.

7.2.2 The Automotive Industry

CAD/CAM has been used in the automotive industry since the early 1960s. It is now employed by car, truck, coach, train, crane, motorcycle, and locomotive manufacturers as well as by suppliers of all sorts of parts such as tires, windscreens, headlights, fuel sensors, seats, spoilers, and engines.

Within the overall design and manufacturing engineering development cycle of a car, CAD/CAM can come into play as early as the styling stage. The geometry of the car body can be defined by the system, and the car's appearance under different color and lighting conditions examined. In 1985, the majority of car manufacturers still produce a full-scale clay model of the car during the design engineering process. The clay model is digitized and the digitized points entered into the system's data base. They form the basis of the definition of the exact overall body surface geometry. This application is carried out with a 3D complex surface geometrical modeller. Special functions to analyze curve and surface properties such as curvature are used to ensure that surfaces are aesthetically pleasing and smooth. Once the complete body surface is believed to have been correctly defined, the geometry model is generally checked by using it to NC machine a model of the car body in soft material. Once the external body shape has been defined, the interior body panels can be defined. CAD/CAM geometry data is input into programs for structural analysis (e.g., to check the strength of a panel), kinematic analysis (e.g., to check the action of a door hinge), aerodynamic analysis, visibility analysis (e.g., windows and rear mirrors), and space allocation (e.g., luggage, engine, and passenger compartment layouts).

The geometry data is transferred to the manufacturing engineers who design and manufacture the press tools for the body parts. The die must contain a cavity shaped like the body part. In addition, the surfaces surrounding the cavity, which control the sheet metal as it is punched, have to be designed.

CAD/CAM techniques are also applied for the powertrain (engine and transmission). Many major suppliers to the car industry employ CAD/CAM, in particular for the detailed design of parts that have been broadly specified by the car manufacturer.

In the manufacturing engineering area, it has been seen that CAD/CAM techniques are put into practice to design body panel stamping dies. They are also used to assist NC programming of dies and moulds for individual parts, to program (or to check the actions of) welding, painting, and assembly robots, and to design and manufacture fixtures and tools for car body assembly.

7.2.3 The Ship Industry

Ship hull design was one of the earliest applications of CAD/CAM and has been applied to tankers, yachts, military surface vessels, and submarines. The process today has been extended to handle outfitting, piping, and other systems. CAD/CAM is now used at the conceptual design stage, the detailed design stage and in preparing numerical control tapes. It is used by suppliers of parts as diverse as pumps, propellers, engines, and furniture.

At the preliminary design stage, the hull can be designed in agreement with the relevant national and international rules and regulations, and then analyzed to give mass and volume properties. Once the overall internal structure including tanks and cargo compartments has been laid out, CAD/CAM geometry is input to static and hydrodynamic analyses. It is possible to evaluate stability under different loading or damage conditions.

At the detailed design stage, CAD/CAM is used to plan the most economical cutting pattern for the ship plates and to produce the corresponding numerical control tapes for the flame cutters. CAD/CAM is also used for the detailed design of the other structural elements (bulkheads, frames, etc.), and for the exact layout of the power, electrical, heating and associated systems. Standard parts libraries containing stiffeners, notches, brackets, valves, and pipe fittings are used extensively. In many cases, CAD/CAM is invaluable in designing complex systems that must be squeezed into confined spaces and producing the corresponding isometric drawings that will be used in manufacture and assembly. Piping can be designed with the system and the resulting geometry used in producing tapes for numerically controlled pipe fabrication.

7.2.4 The Offshore Engineering Industry

CAD/CAM techniques were put into practice in the offshore industry in the mid-1970s. In the conceptual and pre-engineering design phases they are used to build three-dimensional models of rig and platform structures. Structural analysis is carried out on these models to ensure that they meet international standards. The major applications of CAD/CAM in the detailed design phase concern structural steel and the various support systems.

In the structural steel area, the CAD/CAM system is used to calculate weights and center of gravity data, to generate construction drawings and materials lists, and to produce numerical control tapes for plate-cutting machines.

In the support system area, the CAD/CAM system is used in the layout of electrical wiring, the power systems, and the different types of piping. Manual pipe design is a complex and labor-intensive process. Productivity in this area can be greatly increased by employing CAD/CAM techniques. Within the system, a three-dimensional model of the various piping systems (hydraulic, ventilation, air-conditioning, fresh-water etc.) is built and then checked to ensure that there is no interference between the pipes (or structural parts) and to ensure that sufficient clearance exists. Additional software can be used to calculate pressure and temperature distributions, and noise levels in ducts. Standardized pipes and associated equipment stored in a parts library can be used to reduce the need for redesign and to assist attempts to increase standardization. Many piping networks contain repeated common subnetworks. Once one of these has been designed, it can be reused throughout the network. Once a correct model exists, it can be used to produce isometric working drawings, the bill of materials and numerical control programs for pipe cutting and bending. Hydraulic power and control systems containing standard components such as pumps, valves, and regulators can be designed and manufactured more productively. Where manifolds are used to interconnect the components, the complex layout of pathways connecting the various inlet and outlet ports can be designed to avoid interferences and maintain necessary wall thicknesses.

7.2.5 The Machine Tool Industry

Within the machine tool industry, CAD/CAM is used in engineering both the machines themselves and the tools.

Even in companies building special, one-off machines, there is often a common core of major assemblies used in different machines. Costing and design of new machines can benefit from reuse of standard parts (spindles, bearings, foundations, etc.) stored in a parts library. All sorts of drawings ranging from layout drawings of complete machines down to detail drawings of special toolholders can be produced quickly for proposals. CAD/CAM is also used to design the correct gear sizes and combinations to connect the spindle to the motor. Assistance can be given to the designer concerning the type of gearing required (helical or spur), and the suitability of given tooth numbers and materials. Kinematic simulation will be carried out with the system to ensure that there will be no interference between moving parts. CAD/CAM is used to produce assembly and detail mechanical drawings and associated parts lists, as well as electrical wiring diagrams, panel layouts and their associated parts lists. A further major use of CAD/CAM is in the generation of exploded views, and other electrical and mechanical drawings for technical publications. It is also employed by manufacturing engineers in tool and fixture design and manufacture, in the production of NC tapes and in the production of process plans.

Machine tool manufacturers involved in supplying complete systems such as flexible machining cells including automatic material handling devices use CAD/CAM for generating plant layouts. Standard machines and devices can be designed once and stored in a parts library. Then as a function of the plant size and shape, a suitable plant layout can be designed. This can be analyzed on the screen and if necessary modified, thus avoiding the need to correct errors after construction. Simulation of the movements of transfer devices can also lead to improved design before implementation.

Although the tools (e.g., milling cutters) are often specific to a particular product's requirements, in many cases it will be quicker to modify an existing tool than to design a new tool from scratch. It is therefore important for tool manufacturers to have a complete library of existing tools within the CAD/CAM system. When a tool is required for a new product, the designer first checks to see if any of the existing tools are suitable without modification. If not, a design that is similar to that required may be modified on the screen and then passed to an analysis program to carry out the extensive calculations required to ensure that the tool will meet requirements. The system can then be used to produce the necessary drawings.

7.2.6 The Footwear Industry

CAD/CAM is used at many stages of the design and manufacturing engineering processes in the footwear industry. At the marketing stage, shoes can be designed on the screen and then photographed in different colors, from different angles and under different lighting conditions. The photographs can be used to test customer reaction—without any investment in production. Shoe lasts, which have very complex surfaces can be designed with the system, and the corresponding numerical control tapes produced. The shoe upper (a complex surface, three-dimensional shape) can be designed in the system, and then the corresponding flat shape calculated. Cloth uppers are often cut by numerically controlled water jet and laser cutters which can be programmed with data from the design phase. In other cases, cardboard patterns corresponding to the shapes of leather needed to produce the upper can be cut out, and used in designing the corresponding leather cutting hand tools. Generally only one size of a model of shoe will be designed by the designer, the other sizes will be automatically calculated by a technique known as grading. The moulds for plastic soles can be designed with CAD/CAM and then NC machined. CAD/CAM is also employed in designing moulds for plastic ski boots and in designing decorative stitching on boots.

7.2.7 The Power Generation Industry

CAD/CAM has been used since the 1970s for a variety of turbine, pump, and compressor applications. Among the reasons for the early use of CAD/CAM in these areas are the strict mathematical definition and high precision required of some parts of the machinery. These requirements are often customer specific and, with manual methods, lead to a long cycle time of many months. Applications in these areas are heavy users of computer-based structural analysis methods, for which a complete geometric model of parts is required. Since a large proportion of existing parts will be used, designers can benefit from the use of parts stored in parts libraries.

Turbine blades themselves were among the first parts to be handled with CAD/CAM techniques. Blade sections meeting the designer's requirements are used to build up a complex surface three-dimensional representation of the blade surface and root. This is then blended to the rotor wheel. The geometry of the resulting surfaces is meshed and the node points are input to structural analysis programs. The geometry of the blade can also be used in the preparation of numerical control tapes for machining prototype and low quantity blades or for machining blade dies or electrodes.

Apart from blade applications, CAD/CAM is also used in designing piping layouts and sheet-metal casings, as well as in ensuring that there is no interference between parts and in producing detail drawings. In some cases, the design process has been integrated with automated process planning systems.

7.2.8   The Clothing Industry

CAD/CAM is used in the design of clothes, in optimizing the material cutting process to ensure that as little material as possible is wasted, and to produce numerical control tapes to drive knife and laser fabric cutting systems. It is also used to control decorative sewing machines and in preparing knitting machine control programs. Another area in which CAD/CAM has been applied is the design and manufacture of embroidery.

7.2.9   The Hand Tool Industry

Among the major reasons for the use of CAD/CAM in the hand tool industry are the high number of similar hand tools in a family (e.g., a set of spanners) and the high number of fairly similar working tools that is required in the manufacture of a given tool.

Once a given member of a family of hand tools has been parametrically designed and parameterized in the CAD/CAM system, it is not only possible to produce the necessary drawings and the required numerical control tapes for that tool, but also, by changing the parameters, to design very quickly the other members of the family.

Similarly, many of the working tools have a family resemblance thus once one has been parametrically designed, the others can be designed very quickly. The working tools are often built up from many standard parts with only a few parts being specific to the actual tool to be manufactured. Time and cost savings result if the designer can access standard parts stored in a parts library.

7.2.10 The Household Appliance Industry

In today's consumer society, household appliances (vacuum cleaners, telephones, television sets, shavers, hair dryers, etc.) are under continual modification to meet the requirements of the marketing department. One way to meet the need to reduce drastically development cycles has been the use of CAD/CAM. At the conceptual design stage, it is possible to test out several variants and decide which would best meet the expected market requirements. Detailed design will then be carried out with the system and numerical control tapes produced for moulded parts.

7.2.11 The Packaging Industry

CAD/CAM is applied to the manufacture of many types of glass and plastic container (e.g., shampoo bottles, perfume vials, liquid detergent bottles, drinking glasses, alcoholic drink bottles). Due to the need within the consumer society to attract clients with something that is fashionable today but will be outmoded in a few months, there is a continual need to redesign containers. With traditional techniques it is often difficult and time consuming to modify the design of the external aesthetics of a container while maintaining a constant internal volume. This problem is greatly reduced with CAD/CAM.

At the conceptual design stage, CAD/CAM will be used to model several possible geometries. Color shading and varying light source techniques will be applied to show how a particular design would appear on a shop shelf. Geometry information is also used to calculate the internal volume of the container, and to calculate the weight of material required to make the container. Structural analysis of the container will show its resistance to the type of shocks that it is liable to come up against. Once the design has been accepted, detailed drawings can be produced on the CAD/CAM system and the corresponding moulds designed. The numerical control program to produce the mould can then be developed.

7.2.12 The Machinery Industry

The machinery manufacturing industry has a wide range of clients ranging from users of papermaking machinery to users of packaging machinery. CAD/CAM is of particular use in the design of machinery due to the relative similarity of machines made for different clients. The use of standard parts libraries is therefore of a high importance. Once a part has been designed for one machine it can often be reused without modification, or with only minor modifications on other machines. Parametric design can be carried out for parts that have slightly different sizes but nevertheless belong to frequently used, well-identified families. Mechanisms can be designed with CAD/CAM and complex linkages examined to ensure that they will not cause clashes. High tolerance parts can be designed to fit at the first attempt. Assembly and detail mechanical drawings and associated part lists can be produced with CAD/CAM, as can electrical wiring diagrams, panel layouts and their associated parts lists. Exploded views, isometric drawings, and other electrical and mechanical drawings can be produced for technical manuals.

In manufacturing engineering, sheet metal parts can be nested as a function of existing plate sizes. Once the most efficient nest has been designed, control tapes can be prepared for NC machines. The NC programs for parts that have to be turned or milled can also be developed directly from design geometry by manufacturing engineers.

7.2.13 The Electromechanical Industry

Electromechanical components are used in a wide range of industries. These components include electrical fittings, plugs, switches, transducers, switchboards, measurement instrumentation, process control panels, and computer peripherals such as disk units and pen plotters. In many cases, such components can be characterized as small, high quality, and high precision. There may be many moving parts that must fit closely together, and the whole component may need to be built to tight tolerances. It is therefore important that product design be of the highest quality and CAD/CAM can play a major role in ensuring this. A library of standard parts often serves as the basis of the design process, with new parts being designed and fitted in to complex layouts within restricted volumes. Once such a design has been completed, it is a great advantage to be able to visualize the object in three dimensions on the graphics screen, rather than having to wait for a prototype to be built. Once the design is finished, CAD/CAM will be used to produce assembly and detail mechanical drawings, and electrical schematic drawings. Manufacturing engineers may make use of CAD/CAM to nest sheet metal parts in preparation for production by NC machine, or for the preparation of plastic injection moulds.

7.3 CURRENT USE OF CAD/CAM IN VARIOUS MANUFACTURING TECHNOLOGIES

7.3.1 Forging

Although the forging process itself is not changed by the use of CAD/CAM methods, these offer opportunities to improve greatly the quality and speed of design and manufacture of forged parts and to reduce time cycles.

Preparation of the various forging dies by conventional methods is lengthy and based largely on empirical methods. It involves several steps. Starting from knowledge of the geometry of the required part and of the forging conditions, the final forged part geometry is calculated. From this, the finisher dies and the blocker dies will be designed, and initial billet requirements estimated. The dies themselves can be manufactured by conventional die sinking, by copy milling a model, by numerically controlled machining of the die block or by EDM (electrodischarge machining) techniques.

When CAD/CAM techniques are introduced to forging, the initial task with the system involves modelling the geometry of the required part. The geometry is either modelled as a set of two-dimensional cross-sections along the length of the part, or in three dimensions. Addition of appropriate information (e.g., draft angles) needed for the forging process allows the final forged part geometry to be defined. Preliminary design of the finisher die (including flash dimensions) is then carried out and the resulting geometry used as input to analysis programs which, for given forging conditions, will calculate temperature and pressure distributions throughout the die. Calculations can also be made to analyze how the die cavity will be filled. The results of the analysis program may show that the preliminary design has to be modified. After modification to the geometry, the process can be repeated until a satisfactory finisher die geometry is available.

From the finisher die geometry, a preliminary design of the blocker die geometry may be carried out. Using this and knowledge of the forging process parameters (temperatures, materials, etc.) and of the preformed billet shape, the flow of metal in the blocker die can be simulated in the computer. The geometry of the die is modified until a suitable design exists. CAD/CAM techniques can also be used to assist in calculations for performing activities carried out on the initial billet prior to the blocker operation. Once the geometries of the dies have been completely defined in the system, the NC cutter paths can be programmed directly and checked at the screen. Depending on the number of dies to be produced, the die geometry and material, and the availability of machine tools, it may be decided either to NC machine the dies directly in the die block or to NC machine graphite electrodes and then manufacture the dies by EDM techniques.

CAD/CAM has also been used in the design and manufacture of the exact shape of each set of dies needed to attain the final shape of a part to be produced by roll forging.

When selecting a CAD/CAM system for forging it is particularly important to consider the type of geometry of the parts made (from simple uniform cross-sections to singly or doubly curved surfaces), the complexity of the geometry (holes, bumps, etc.), the number of parts designed each year, the number of dies designed, modified, and recut each year, the possibilities of parametric design for parts in the same family, etc. A decision must be taken as to whether the system will be used in analysis, geometry modelling, and NC tape preparation or just in drafting (i.e., in producing the same part, die, and fixture drawings currently drawn by hand).

7.3.2 Plastic Moulding

Among the major requirements for making plastic parts, for example by injection moulding, are the needs for mouldable design of the part and a correctly designed mould. By conventional methods, mould design is very much an experience-based skill. Design of a mould involves much more than design of the mould cavity since consideration must also be given to the optimal number of cavities and their positions, the number, shape and position of gates, the clamping, cooling, ventilation, and ejection systems, etc. Without computer-aided methods, design cycles are often very long and the quality of the moulded part is far from optimal.

A typical part to be moulded might first be styled to meet consumer requirements. The material would be chosen and the manufacturing technologies defined. Preliminary design would then be carried out to define and position the external and internal arrangements of the part. Each of the individual structural elements would be designed to ensure its suitability. By conventional techniques the process is very time consuming and generally requires a prototype mould to be designed and prototype parts produced.

The first stage in the CAD/CAM-assisted process would be to model the geometry of the part. The system could be used to check the volume and center of gravity of the part. The part could then be analyzed by a computer program to ensure that it is sufficiently strong or that it can be easily opened. Analysis may show that a part has been overdesigned. Once a suitable part geometry has been defined, the system can be used to produce any required part drawings.

From the part geometry, the mould cavity geometry can be defined. This may involve use of the system to calculate shrinkages. The mould maker can use standard objects (screws, slides, hydraulic cylinders, ejector pins, dowels, mould bases, etc.) available in a parts library to assist in design of the mould. This reduces the need for redesign and repetitive detailing. Once the complete mould has been designed, analysis programs can be used to check its suitability. Programs are available to simulate the flow and cooling of plastic through the mould. The effects of different runner sizes and gate positions can be investigated. Part and mould cooling time can be calculated. The preliminary design can be modified until the simulated flow and cooling time become suitable. Then the system can be used to produce all necessary drawings and the bill of materials. The CAD/CAM system can also be used in deriving the NC tool path for complex mould cavities and other mould components. The tool path can be checked on the screen.

7.3.3 Numerically Controlled Machining

Numerically controlled machines use data rather than humans to control the machine tool movements that result in a part being machined. Much of the data defines required axis movements, other data defines spindle speeds, feed rates, coolant selection, and other such factors. The input to a NC machine tool (i.e., the NC program) has traditionally been on punched tape. NC is used for a wide variety of machining ranging from drilling, boring, tapping, and turning to 3-axis and 5-axis milling. NC is also used to control other tools such as flame cutters and fabric cutting knives.

Traditionally, an NC programmer (or part programmer) has developed the NC program from part drawings. Development has been either manual or computer assisted. NC languages (e.g., APT) have been introduced to ease the preparation of NC programs.

Within the CAD/CAM context, NC has taken a great step forward. Instead of picking up the part geometry from a drawing, the part programmer can now use the geometry in the CAD/CAM data base as modelled by the part designer. This leads to a saving in time and a reduction in transcription errors. The part programmer can then select a tool, define the part to be machined and how it is to be machined. The machine table geometry can be retrieved from a parts library along with other standard parts such as clamps and pins. The part programmer defines the fixture base and the positions of the fixtures. Depending on the type of CAD/CAM system, the type of NC machine being used and the part geometry, the system may then calculate the tool path. In general, the simpler the part and the machine, the more chance there is of the system being able to "automatically" generate the tool path. Conversely, on a 5-axis machine and doubly curved part, the system will be less likely to be able to automatically generate the tool path. Some systems are capable of automatically preparing the path for such operations as pocketing and threading, and can automatically produce a roughing path to remove stock from a blank. Once the tool path has been defined it can be simulated on the graphics screen, and tests may be carried out to ensure that the tool will not collide with the part, a clamp or the table. This simulation is of tremendous benefit as it reduces nonproductive time spent on "prove-out."

In general, NC machines and their controllers have series-specific specifications. Thus, once the cutter path and machining parameter data have been defined, this data is passed through a postprocessor specific to a given NC machine, and the data to control that machine is output. This output may be on paper tape, or it may be transmitted directly by computer to the machine tool. The use of one CPU to control one machine tool is known as CNC (computer numerical control). The control by one CPU of more than one machine tool is known as DNC (direct numerical control).

7.3.4 Sheet-Metal Working

Flat sheet-metal shapes can be nested interactively to optimize raw material utilization and reduce scrap. Once the shapes have been nested on the graphics screen, CAD/CAM techniques can be used to produce the control program used to drive the NC machine that will cut out the shapes. Compared to the manual, traditional process, major gains result from increased material utilization and from the fact that templates are no longer needed.

CAD/CAM techniques can be used to develop (or flatten or unfold) a three-dimensional sheet metal part into its corresponding flat pattern shape. This greatly reduces the need for trial-and-error methods in determining the exact prefolding shape. With CAD/CAM, automatic calculations to determine the exact bending and stretching are carried out as a function of the type of material, material thickness, and bend radius. Drawings of different stages of the bending process can be produced to assist prefabrication checks.

Another area of sheet-metal working in which CAD/CAM can be of assistance is in the design of sheet metal dies (e.g., for car body panels). The manufacturing engineer retrieves the shape of the finished part from the CAD/CAM data base, and specifies tooling in the regions outside of the trim line to control the metal shape during the punching process. (Some analysis methods, which make use of user-supplied data such as press and part parameters, and type, shape, and size of the corresponding blank sheet, have been developed to calculate the blank sheet required for a particular die. However, much more work must be carried out before such techniques can be used automatically.)

7.4 SOME NATIONAL AND INTERNATIONAL CAD/CAM PROJECTS

In many countries in the world, governments and industrial organizations support research and development of CAD/CAM and the introduction and use of CAD/CAM into industry. Some of the earliest, largest, and best-known schemes were funded by the U.S. military. However, funding has also been made available (directly or indirectly) in Europe (e.g., France, United Kingdom, West Germany, Switzerland), Southeast Asia (e.g., Japan, South

Korea) and Australia. The United Nations Industrial Development Organization (UNIDO) has assisted developing countries with schemes to promote the use of CAD/CAM. Some of the European Economic Community's ESPRIT projects are concerned with CAD/CAM. Computer manufacturers have given grants and equipment to universities to establish CAD/CAM facilities.

In the United States, NASA's IPAD (Integrated Program for Aerospace Vehicle Design) was launched in 1972. By 1981 it had progressed to the stage at which the basic requirements for an integrated CAD/CAM system and the key elements of the system had been defined. In the system envisioned, the individual engineering and mar.ufacturing activities (performance specifications, conceptual design, detailed design, drafting, manufacturing, and inspection) are integrated through a common data base. The primary engineering interface would be through interactive graphics terminals to enable definition and display of designs, and to select and control events. The four key technological ingredients required for an integrated CAD/CAM system were found to be the ability to manage and access data, the capability to distribute data over a network of computer systems from different vendors, the availability of part geometry, and the provision of software utilities to assist interfacing to the system. Along with the above technological requirements, IPAD also noted that the success of an integrated CAD/CAM system is heavily dependent on management commitment and support.

From 1977 onwards, the U.S. Air Force ran the ICAM (Integrated Computer-Aided Manufacturing) program for the U.S. Department of Defense. The progrem, involving U.S. government agencies, private companies, universities, and research institutes had a total budget of $100 million. Mainly directed toward manufacturing of discrete parts in small batches, its aim was to establish modular subsystems that may be computerized and integrated together. It has been shown that the most critical problems facing manufacturing companies trying to improve productivity with computer-based techniques, include poor system integration, poor data quality and accessibility, and system inflexibility. It is proposed that manufacturers wanting to use integrated computer-aided manufacturing technologies should take a top-down approach and first define an overall architecture and the requirements for common versus private data. Individual applications can then be automated using the architecture and the data requirements as top-down controls. The ICAM program was followed up by the CIM program.

In 1972 CAM-I (Computer-Aided Manufacturing International, Inc.), a nonprofit organization was set up to further the research and development efforts of companies with common interests in the CAD/CAM area. CAM-I is member owned and draws its support from governmental, industrial, and educational organizations in many countries. Member organizations can participate in projects and committees aimed at meeting industry's need for computer systems and software capable of increasing productivity and improving the quality of working life.

7.5 CONCLUSION

It can be seen that companies in the mechanical engineering and manufacturing engineering sectors should not still be asking themselves the question "Should we use CAD/CAM?". Experience drawn from different industries in different countries shows that CAD/CAM can be used productively throughout the design engineering and manufacturing engineering process. It is now very important for companies not only to use CAD/CAM, but to use it as efficiently as possible. There should therefore be a slight move away from the purely technical approach to CAD/CAM (in which CAD/CAM is judged by the functions available in a particular system) to a more management-oriented approach (in which CAD/CAM is judged by its overall effect on a particular company).

8

INDUSTRIAL CASE STUDIES

The following case studies describe the choice, implementation, and use of CAD/CAM by a variety of companies. These companies use a variety of systems to handle a wide range of industrial applications. Some of the companies are large, others are small. The case studies are intended to be factual, and no criticism of particular CAD/CAM systems or the way in which these systems are used by the companies is intended.

The assistance offered by the companies involved is gratefully acknowledged. In particular, the following people are thanked: R. Addor, S. Alber, P. Bean, F. Busset, W. Dankwort, J. F. Doray,

A. C. Henton, M. Hugi, J. Kolarik, P. Lachmann, A. McLean,
J. Norton, G. Oswald, R. Schuster, D. Trebley, and I. Williamson.

8.1 APV INTERNATIONAL LTD

APV International Ltd of Crawley, West Sussex, England is a
member of the APV Group. There are about 1200 employees on
the Crawley site where equipment such as stainless steel vessels,
tanks, pipes, and plate-type heat exchangers are designed,
engineered, and manufactured for breweries, dairies, and chemical
plants. Automatic process control installations are also developed.

APV first used computers in a technical environment in 1964.
In 1970, it was recognized within the company that computer
graphics was going to be an important tool for improving design
productivity. An investigation at that time showed that the most
productive application of CAD would be for the design and layout
of piping. As a suitable program could not be found on the
market, it was decided to develop a CAD program in-house. The
program, written in FORTRAN, became known as Pipework.
Initially it was developed to run on a bureau computer and then
on a Prime 300 computer (with 2 1.2 MB cartridge disks) and
Tektronix terminals. The program allows the designer to choose
and layout pipes and standard fittings. It checks automatically
that clashes do not occur and that sufficient room is available
around pipes and fittings for manipulation. Another feature is
the automatic insertion of necessary junctions into long pipe runs.
For each national or international standard required, a set of
about 200 commonly used, standardized pipes and fittings is built
up and stored in a library. Use of these parts reduces design errors,
enforces standardization of parts, ensures that standards are
adhered to, and reduces costs and time cycles in the overall design
to assembly process.

The use of Pipework has led to benefits in tendering and in the
Drawing Office. Since drawings can be produced faster with CAD,
the response time for quotations has been reduced. The modifications and extra views requested by prospective clients have been

produced much faster. In some cases, clients have only placed the final order on the condition that all design work will be carried out with CAD. In the Drawing Office the major gain resulting from the use of Pipework stems from the much greater speed and accuracy with which drawings are produced. In particular when small changes have been made to a design, the modified drawings with corresponding changes have been produced much more quickly than would have been possible by manual methods.

The Pipework program contains a feature called *bulk issue* to estimate the overall material quantity for a particular contract. Before the detailed engineering of a contract is started, this estimate is passed to the Stores. Manufacturing staff can then even out the workload for the contract rather than carrying out all work once the design has been finished. As the detailed engineering continues, the program matches quantities against estimates and advises Stores of any differences.

Pipework provides automatic material takeoff lists. This feature has been found to significantly reduce costs. Since the lists are 100% accurate, neither underordering nor purchase of surplus parts occurs. When changes have been made to a design, the modified material takeoff lists have been produced quickly and accurately.

Accurate isometric drawings produced with Pipework have led to major productivity gains at construction time. With manual techniques, it was often found during construction that pipes did not fit properly, and pipe fitting was generally left to the very last moment of plant construction, by which time exact positions were known. With CAD however, the confidence level in assembly drawings has been much higher and sometimes prefabricated pipe runs have been made up offsite. Assembly has also been made easier because of upstream improvements resulting from CAD, such as improved design checks and accurate material takeoff lists.

In mid-1985, the Pipework program was being run on a Prime 500 computer (with 2 300 MB disks). It had taken about 10 man-years of development work, and ran in about 600 KB. Six workstations were dedicated to the system's 12 users who used it on a one-shift basis. Pipework is supplied to the users in the Drawing Office by one section of the Technical Computing Group. Its everyday availability is the responsibility of another section of this group. A user of the system (in fact a Drawing Office Section leader) has been designated as System Administrator and is responsible for such activities as scheduling system usage time, running the terminal room, training, setting standards, and producing documentation for the users. Training for a new user consists of 2–3 days basic training after which guidance is given as required for about a month.

Since 1975, Pipework has been used profitably on more than 200 projects. The projects tend to be very different. Project data is regularly archived on disk and then to magnetic tape.

A program was developed in 1981 and is now in everyday use to thermally design and analyze spiral heat exchangers. At the proposal stage it produces a cost estimate and the proposal drawings including the parts list and heat-exchange parameters. Once the contract is signed it is used to produce manufacturing planning and detailed drawings.

In 1982, it was decided to extend the use of CAD to the design and drawing of standard components for manufacture such as valves, pumps, and stainless steel fittings. There are thousands of these parts and their costs had to be reduced if the overall manufacturing costs were to be reduced. In 1982, Pipework and technical calculations were being run on a Prime 500. A 6 month trial of DOGS (from Pafec) running on the Prime 500 was carried out simultaneously with a study of CAD/CAM requirements and solutions. As a result it was decided to buy DOGS on a Prime 250 with five workstations. In mid-1985, the system was expanded by the addition of a Prime 2250 and two more workstations. At that time, the six Pipework workstations had about 12 users and the seven DOGS workstations had about 20 users. The three computers have all been installed in the computer room and are networked together. In contrast, the terminals are concentrated in a Pipework terminal room and a DOGS terminal room. The DOGS terminals have been used on a two-shift basis from 07.00 to 16.00 and from 10.00 to 18.30. The amount of time that a user spends in front of a terminal varies from one to four hours per day. Initial DOGS training usually takes place at the Pafec site with further training being given in-house. When a new version is installed, a senior user follows a course at Pafec and then passes on the information to the other users. As with Pipework, the responsibility for everyday availability of DOGS lies with the Technical Computing Group, while a System Administrator drawn from the users is responsible for its smooth running. Whereas a lot of initial ideas concerning CAD/CAM have come from the Technical Computing Group, actual responsibility for setting the CAD/CAM strategy lies with the Drawing Office manager.

The use of DOGS has led to a major increase in the speed of drafting. It is especially appreciated for the gains that occur when minor modifications have to be made to a drawing and when existing or standard parts can be reused. Major time savings occur when it is used for line diagrams. It is also found to be very useful in redesigning parts so that they are less expensive, of higher quality, and make optimum use of a particular machine's capabilities.

APV is confident that profits have been made at every stage in the design and manufacturing process to which CAD/CAM has been introduced. Management sees CAD/CAM as an important tool for improving productivity, in particular for cutting costs in the manufacturing area. APV was one of the earliest engineering companies to make use of computers for technical calculations and for CAD/CAM. APV continues to invest in the development of its CAD/CAM resources. The next major steps forward foreseen are to integrate CAD/CAM with technical calculations, and to integrate CAD/CAM with manufacturing resource planning (MRP).

In the early 1970s it was felt that CAD/CAM would offer "push-button" design. However, although CAD/CAM has automated some parts of the design process (e.g., technical calculations and graphics), it is still a long way from automating the experience- and knowledge-based tasks and the necessary interdesigner communications that make up such a large part of the design process.

8.2 BMW AG

BMW AG had about 44,000 employees and a turnover of DM 11,480 million in 1983. CAD/CAM is used in both the automobile and motorcycle businesses. The major objectives behind the initial decision to invest in CAD/CAM (taken in 1977) were to reduce the overall design and development time cycle, to increase productivity by integrating previously independent phases of this cycle, and to optimize design. The first CAD/CAM system was in place in 1978. By 1980, 20 workstations were in place, and by the end of 1985, about 200 CAD/CAM workstations had been installed.

The major phases in the design and development cycle of a car are predevelopment, concept development, detailed design, prototype manufacture and testing, design of tools and fixtures for production, manufacture of these tools and fixtures, planning and quality control, production of the preseries and finally, series production. Traditionally (i.e., before CAD/CAM was used), each phase was carried out separately and in sequence, and little work could be carried out on a phase until the previous phase had been terminated. With CAD/CAM it has been possible to increase the overlap of design phases and, for example, to pass data from the first phase to the second before the first phase has been completed. Similarly, new opportunities arising from the use of 3-D representations with CAD/CAM allow the designer to carry out stress, kinematic, collision, and assembly analysis before a prototype has been built. This not only saves time but leads to design optimization.

It was recognized that different application areas have different CAD/CAM requirements. At the time at which CAD/CAM was introduced, no individual CAD/CAM system met all these requirements. The solution of a single CAD/CAM system for all applications was rejected as being unproductive. Such a system would, in general, only have been used in carrying out the most mundane parts of the design. At that time, had such a system been chosen, it would only have been of assistance in drafting. This solution would not have met the objectives mentioned above, nor contributed much to the overall product requirements such as high quality, high precision, attractive design. It was therefore decided that the best possible system for an application (i.e., the system meeting as closely as possible the particular application's requirements) would be applied to that application. This led to the use of several CAD/CAM systems within the company. Each of these systems was required to meet the specific requirements of the application for which it was used. The order in which systems have been installed reflects the benefits expected to arise from their use in a particular application area.

CAD/CAM systems were first installed (in 1978 and 1979) to support car body applications as this was the area believed to offer the highest potential productivity gains. Within this area, the individual activities include styling, model manufacturing, digitization of models, production of computer-based model drawings, smoothing of surfaces, designing, model generation by NC milling, tool designing, tool manufacturing (copy milling), and checking. With CAD/CAM it has been found possible to create more body design alternatives within a reduced time, and to increase the quality of the body. The systems used for body applications are GILDAS, MEFISTO, and STRIM. GILDAS is an in-house development for managing the multitude of digitized points produced from models. MEFISTO is another in-house development. It is a surface milling system with 5-axis capabilities. STRIM (from Cisigraph), a surface modeller, based on a multiparametric polynomial representation, is used by the designer to 'smooth' the digitized points to form individual patches of surface. These patches are then blended together, and modified if necessary, to form an aesthetically pleasing car body surface. All of the outer body surface is designed using STRIM. The same system handles many of the inner body parts. Although many of these are not made up of such complex shapes as the external body, they are often designed using information available in the external body description (e.g., offset surfaces). The system is also used to design interior fittings and mouldings for parts such as seats and sun visors. It is also used in windscreen design and manufacture; for example to design a developable windscreen surface to fit the requested windscreen outline, or to calculate the best shape of the flat glass that will be moulded to produce a nondevelopable windscreen surface.

Once the use of CAD/CAM had been successfully demonstrated in car body applications, CAD/CAM systems were implemented for other applications. In 1979, CABLOS (from AGS) was implemented for schematic diagrams and layouts. In 1980, CD-2000 (from CDC) was implemented for the design and drafting of mechanical parts. CADAM (from Lockheed/IBM) and CATIA (from Dassault Systemes/IBM) were then installed for manufacturing engineering applications such as design of press tools, casts, fixtures and production machine mechanisms, and for preparation of numerically controlled machine tool programs. Finally, 1983 saw the development of GRIVAD, a system for circuit design, electrical wiring layout and electrical parts list generation.

BMW have purchased CAD/CAM systems wherever possible (i.e., whenever there has been a system available on the market to meet the requirements of a particular application). In-house developments have been made when a suitable system has not been found on the market. Typical in-house developments have been in linking systems together, and in special car industry-related applications (e.g., kinematic analysis of wheel movements, calculation of visible areas, and calculation of the wiped area on the windscreen).

It has been found that the initial acceptance of a CAD/CAM system is decisive for its long-term success. Another requirement for success is a constructive dialogue between system managers and users, with the users being able to positively influence the development of the system. Other important requirements have been found to be high stability and availability of the system, transparency to the user of EDP problems, and the possibility to adapt the system to specific requirements of the company (particularly with respect to data interfaces, data protection, and special applications).

BMW have found that the use of CAD/CAM leads to savings in time, to lower costs, to higher flexibility, and to increased product quality. It also offers, in some cases, the possibility to carry out tests that were just not possible before introduction of CAD/CAM. Reduction of time cycles is particularly appreciated partly because it offers the possibility to create alternative designs within a given time period, and partly because it offers, for example, at early stages of styling and design, the chance to reduce lead times. Quality improvement both produces a better product and results in a reduction of harmonizing and modification work at later stages of the manufacturing process.

The period 1979 to 1984 is seen as a highly successful one in which productivity in several application areas was increased by the introduction and use of specific CAD/CAM systems meeting specific application requirements. Since 1983 BMW has been preparing for a new phase of CAD/CAM development in which further productivity gains can be attained by increasing integration between systems. One major requirement is to improve the transfer of CAD/CAM data both between applications and with subcontractors. An in-house development, CADNET, will be used to archive, exchange, and release CAD/CAM data independently of the system and host computer. CADNET uses IGES and VDA formats. BMW is also cooperating with other companies (e.g., on an Esprit project), to attain a unique data interface between systems. Whereas the initial period of CAD/CAM use led to success in specific application areas, productivity gains in the next phase will come both from full integration of systems within the same development phase and from integrating different development phases.

CAD/CAM at BMW is not seen as an isolated technique, but as a major component of CIM. It is therefore developed in conjunction with production automation (NC machine tools, robots, etc.) and communication techniques for improved technical administrative logistics (including process planning and engineering data management).

8.3 A COMPUTER PERIPHERALS MANUFACTURER

In early 1983, a committee was set up within the company to choose a CAD/CAM system for about 40 mechanical and electronic engineers, designers, and draftsmen. The system chosen met the requirements of the committee, ran on hardware the company was familiar with, and contained a suite of software covering electronic and mechanical design and analysis.

In the initial organization, a mechanical system support engineer and an electronic system support engineer reported to the CAD/CAM manager, who reported in turn to the Technical Director. The potential users of the system remained in their traditional design office or drawing office positions.

Of the eight workstations for mechanical design, six were installed in November 1983 and the other two in late 1984. Version 1.002 of the software was installed in November 1983. The company was one of the first to receive this version of the system. At the end of 1983 and early in 1984, 10 users received 8 days training for mechanical design. The system support staff followed an additional 3-day systems manager course.

It was decided to use the CAD/CAM system to speed up the design of a printer then going through the usual design and development process. The design process had started manually one year earlier and significant progress had already been made. However, great difficulty was experienced in using the system features to build an "assembly" made up of several parts. It was also found that the company had no structure to break up a project into manageable parts for individual CAD/CAM users, or to reassemble those parts back into a finished project. The problems became so serious that in March 1984 it was decided to stop using the CAD/CAM system, and continue design of the printer with manual techniques.

From March 1984 to December 1984, the CAD/CAM system was used occasionally to generate isometric views of parts that had been designed manually. These views were then plotted and taken back to the drawing board where they were manually traced over and thus transferred to part drawings. The isometric views increased the understandability of the drawings.

In June 1984, Version 1.5 of the software was installed. This was found to be significantly more stable and it became possible to successfully combine parts together. However, disillusionment with CAD/CAM was running high both among users and the CAD/CAM staff. Before the end of 1984, the CAD/CAM manager, the mechanical system support engineer, and the electronic system support engineer resigned.

In November 1984 it was decided to take a fresh look at CAD/CAM. A new organization was put into place with a PCB system support engineer reporting to the CAD/CAM department manager (who was actually the original mechanical system support engineer), who in turn reported to the Technical Director. The CAD/CAM department manager developed a three-phase CAD/CAM implementation plan which was put into action in January 1985.

In the first phase, two users from the drawing office and the CAD/CAM department manager spent about two months getting to know the system, finding out what it could and could not do. They demonstrated that the system could be used for designing piece parts and for assembling these parts using both wireframe and solid modelling software. The steps involved in creating a detail drawing from a solid model were thoroughly investigated. They also found that the system was vulnerable to people making misguided use of it. During this phase, initial working procedures were put into operation and some procedures were implemented to protect data against illegal access.

In the second phase of implementation, from March 1985 to May 1985, another user joined the small group looking into ways of using the system. During this period, more working procedures were put in place. A dictionary of some 40 tablet symbols was built up and the system was customized (e.g., the dimensioning was set to company standards). Several existing designs were redeveloped on the system to ensure that the same results could be achieved with the system as by manual methods. By this time it was felt that the system was sufficiently user friendly to be used by drawing office staff after an initial training course. However, before the users could be let loose, documentation was produced to help them understand how to use the system. The documentation covered subjects such as part naming conventions, modelling, plotting, rules for dimensioning, account maintenance, archiving and back-up, procedures for handling large projects, and error handling. Work was also carried out with the solid modeller to demonstrate the potential of the system. The results were shown to top management and clearance was given for the third phase to start.

At the start of the third phase, between May 1985 and July 1985, the system was used productively for the first time. Starting from piece parts designed manually, some 30 assembly drawings for the printer were produced. These ranged from very small assemblies up to complete mechanism assemblies containing nearly 40 significant items. The reaction to these drawings was extremely favorable, and for the first time it became apparent to the majority of users and managers that the system was extremely powerful, and if used correctly would be of major benefit to them.

By then, the printer project had moved into the tooling phase and the system was used extensively in design and cost optimization of manufacturing tools, including moulds for plastic injection and dies for aluminum parts. At that time there were five users of the system. Once they had fully understood the system and could use it productively, it was found that they tended to work for long periods at the screen, including during the lunch break and in the evenings.

At the beginning of the third phase, a suite of procedures for managing designs throughout a project was developed, and CAD/ CAM was interfaced to the existing drawing control system.

The following steps were to link the system to ANSYS for structural analysis of printer parts, to interface the CAD/CAM-produced bill of materials (BOM) to the companies' existing BOM system, and to make available a terminal in the manufacturing area for on-line interrogation of models and drawings.

Perhaps the greatest problems with the system, apart from the initial technical ones, have stemmed from the belief that a turnkey CAD/CAM system could carry out "turnkey design." This has been found to be completely untrue in that designers still have to design, projects have to be controlled, data have to be managed, people have to communicate information, and so on. Another problem arose because the workstations were initially in an office environment, and the users were therefore subject to distractions such as telephones ringing, office conversations, and stray beams of light falling on the screen. A purpose-built, controlled-environment area will be built for the workstations. The computer is installed in a special purpose computer room.

Benefits from use of CAD/CAM include overall design cycle time reduction, improved drawing quality, availability of isometric views, faster modifications to existing designs, and the availability of a designed model that can also be used in analysis and in manufacturing engineering. The solid modeller has been of particular use in establishing initial designs and in discussions with suppliers. Since it is possible to show the supplier a well-defined, clearly understandable, computer-generated "solid" image of the part, the supplier can make the part cheaper, faster, and without all the misinterpretations that can arise from deciphering engineering drawings.

Other benefits result from having a single master version of each part. This is used throughout piece-part and assembly work. Any modification to the master is reflected on all drawings where that part is present.

8.4 MARCONI UNDERWATER SYSTEMS LIMITED

Marconi Underwater Systems Limited is a member of the British-based GEC Group. Of the 5000 employees about 1000 are based at the Wembley site in Northwest London. They are involved in electronic and mechanical design and development for underwater systems.

In early 1983, the company's top management decided that the potential of CAD/CAM to increase productivity should be investigated. The major requirement was to reduce the overall design cycle time. This also implied increasing the speed of making customer-requested modifications to designs. In addition, it was hoped that the use of CAD/CAM would lead to a reduction in subcontracting, as this was not only time-consuming due to delays in transmitting and understanding information, but also expensive.

The system chosen was an Applicon system and it was installed in October 1983. Eight workstations were purchased for printed circuit board (PCB) and very large-scale integrated (VLSI) work, and 4 workstations for mechanical design. (All screens chosen were color graphic screens. Color has been found to be very useful, for example in distinguishing between different materials of a part, or between different layers of a circuit.)

To select users for the systems, all staff were first asked to reply to a questionnaire. Only those whose responses were judged to show real motivation to work on the systems and had the ability to use them productively were initially given training. It was decided to build a nucleus of young graduate engineers fully conversant with CAD/CAM techniques. Once this nucleus existed and the major questions of getting to know how to use CAD/CAM productively were solved, more experienced designers were encouraged to use the system. The intention was to have as many design engineers as possible trained to use CAD/CAM.

Some problems arose as a result of using CAD/CAM. For example, on the PCB side some time was lost before a solution was found to the problem of not being able to design large boards. The problem arose because the automatic routing and placement software did not have a true multilayer router, only a double-sided router. It was solved in a later version when a true multilayer router became available. On the mechanical design side, some design engineers were demotivated because the system did not do more of the work for them. They had, in a way, believed that the computer could do almost everything, and they were disappointed to find that it was still necessary for them to carry out the design. However, those who persevered with the system found that it could do all that was required providing that it was approached in the "right" way.

By the end of 1985, there were about ten users of the PCB system and seven users of the mechanical design system. For a variety of reasons, experience with the two systems has been quite different. This was partly due to the more straight-forward character of PCB work compared to mechanical design, and partly because it was much easier to switch relatively small PCB projects to the system. It was not felt practical to introduce CAD/CAM to mechanical projects for which large numbers of drawings already existed. Use of the system for PCB design has been very successful. Once trained on the system, PCB design engineers have used it as a tool to help them in their everyday work. The 8 workstations are used on an overlapping 2-shift basis, and at weekends. In the future, more workstations will be purchased for electronic design and analysis.

The mechanical design system has been organized and managed in the same way as the PCB design system. In both cases, the "system" has been supported by the technical computing support group, and used by users who report to and belong to their original organizational entity—in most cases the Design Office. A design engineer with both engineering and technical computing experience was assigned to the full-time position of mechanical systems supervisor. The major tasks of this supervisor have been to carry out training and supervision of users, to develop special in-house procedures (e.g., for dimensioning, menu layout, data management, and system use), to evaluate new and advanced system features for everyday use, and to try to identify the cause of any problem that may occur during use of the system.

In 1985, it was found that about 9 months were needed for designers to become really competent in using the mechanical design system. They were first given two weeks full-time training. For the next 10 weeks, they spent 50% of their time on productive manual project work and 50% using the CAD/CAM system, under the eye of the system engineer, on "training" projects. Then the users started to use the system on "live" projects, and a further six months' use was required for them to become really competent with the system.

The solid modelling facilities of the system have been widely used. The electronics engineers appreciate the capability to see, in 3D, what a PCB board will really look like. Some designers spend almost 100% of their time at the screen using the solid modeller for mechanical design work. The solid modeller has been found to be of great value in concept modelling where it has been possible to identify and eliminate much more quickly both ambiguities in design and conflicts in assemblies. Other major advantages are in improved communication with clients and potential clients, and in preparing technical and marketing documentation.

It has been found that the solid model gives such a realistic impression of a part that the client finds it much easier to understand the proposed design and ask for any necessary modifications (which may result from previous misinterpretation of requirements). These modifications can be made very quickly and the new design presented to the client within a few hours.

The speed and low cost of producing very understandable conceptual models is one of the major advantages resulting from use of the system. Other gains have occurred because of the ease of making minor modifications to existing designs. It has been estimated that such modifications (and the production of the corresponding documentation) can be carried out about 3 times faster with the system than without it. Installation of the system has led to a reduction in subcontracting.

The major problems that arose from introduction of CAD/CAM were not due to the CAD/CAM system itself but to people. Among the users it was necessary to overcome initial misunderstandings and misconceptions concerning just what the system could and could not do. Then, once trained, users had a much higher market value. Trained users were headhunted by other companies and by agencies, and in some cases left the company. Although top management commitment and assistance were always present and the CAD/CAM organization ran efficiently, it was necessary to overcome initial reluctance among project managers to using CAD/CAM. They tended to prefer to carry out their projects with the manual methods that they were familiar with, rather than taking "risks" with CAD/CAM. To overcome such reluctance it was necessary to demonstrate that CAD/CAM was successful on several real, "live" projects (rather than demonstration projects), and to this end the very visible results of solid modelling have been of great importance. In addition, many briefings have been given to middle management to emphasize the successful use of CAD/CAM. At the senior management level, there has been less of a need to "sell" CAD/CAM, as it is recognized that to stay out in the forefront in a high technology industry, up-to-date, productive systems such as CAD/CAM must be used to ensure survival.

A similar-sized Computervision system, specializing in mechanical engineering design as distinct from electronics, was installed in parallel at Marconi Underwater Systems Limited's Waterlooville Laboratories and has proved quite invaluable.

8.5 PATERSON CANDY INTERNATIONAL LIMITED

Paterson Candy International (PCI) of London, England is a member of the Portals Group. PCI designs, engineers, and commissions water treatment schemes and plants worldwide. Major components of a typical plant include clarifiers, filters, chemical handling and dosing equipment, water and chemical storage tanks, delay chambers, and sludge handling equipment. Data collection equipment, instrumentation, and control panels are designed and engineered to meet specific conditions and requirements. Turnkey developments include all aspects of process, chemical, civil, mechanical, hydraulic, and electrical engineering. Most of PCI's marketing, design, and engineering staff are based in Ealing, London. There are approximately 20 design engineers and 50 design draftsmen. Computers were initially used in PCI for business applications (e.g., financial control) and basic technical calculations.

In early 1983 it was decided to seriously investigate the potential use of CAD/CAM. About 20 CAD/CAM systems were selected for initial investigation, of which 11 were evaluated in depth. Full benchmarks were then carried out on three of these. One person, a project manager, was involved in the entire selection process, with other staff being involved in specific parts of the process. About 10 companies using CAD/CAM were visited. Although several three-day conferences were attended, they were not found to be as useful as a one-day, focused seminar in which speakers from industry told of their actual CAD/CAM experience.

In February 1984, a recommendation was made to the Board of Directors to purchase the DOGS system (from Pafec) on a Prime computer. The major productivity gains expected from CAD/CAM resulted from the reuse of standard parts and computer-based design optimization. Among the reasons for their choice of the particular system were flexibility (the possibility to add or change hardware or software), belief in the future of the vendor (large user base and good development plan), and the quality of the existing system.

The system was installed in August 1984. A 3-phase, 3½-year strategy was developed. During the first 18-month phase, one of the major tasks was to build up the data base. Enough was learned about CAD/CAM during this phase to be able to make detailed planning of the other two phases possible. The project manager mainly involved in system selection was appointed as CAD/CAM System Manager reporting to the Chief Engineer. A design engineer was selected to run the day-to-day operations on a part-time basis. It was found though, that such a position is virtually full time. Various problems arose during the initial period of use, some resulting from connecting together hardware from different manufacturers, and one due to failure of a hardware device.

To familiarize users with keyboards, initial training involved following a self-teaching touch-typing course. Then a single work-station was rented and potential users encouraged to experiment with it and accustom themselves to the system layout, messages, and basic capabilities. Six users were then sent to a three-day training course at the vendor site. Their basic knowledge of system hardware and software allowed them to concentrate on learning about the use of system functions. During the first year, a further five users followed the same course.

The initial system consisted of a 2 MB main memory Prime 2250 with a 68 MB disk, three monochrome screens, and one color screen. It was expanded during the first year by the addition of another graphics screen, a further 2 MB main memory and by upgrading to a 158 MB disk. It was found that design draftsmen need to use the system for about 250 hours over a period of about

4 months to be able to "draw" as quickly with the system as without it. Major gains in productivity were found to come from reuse of standard parts in a parts library, from new designs requiring only minor modifications to existing drawings, from automatic material takeoff, from use of isometric erection drawings, and from linking CAD/CAM to computer-based technical calculations.

8.6 RANK XEROX LIMITED

Rank Xerox Engineering Group at Welwyn Garden City, England, is part of the Xerox Corporation, and its use of CAD/CAM reflects the parent company's CAD/CAM strategy.

The Xerox Corporation had about 100,000 employees in 1984 and a total revenue of about $9000 million. Xerox is well known for its copiers and duplicators, electronic printing systems, memory writers, etc. There are several major families of copiers and duplicators, each of which is made up of five to ten individual products. Each product comes with a wide range of options (paper input systems, stackers, sorters, staplers, color toners, coin operation, etc.) and parts will be made of sheet metal, extrusions, castings, plastics, special elastomers, glass, and include electronic and electrical components. High-quality product engineering is therefore essential, especially as a high proportion of parts is purchased from suppliers.

In the 1970s a variety of systems (including 3 different turnkey systems installed in-house, and service bureau-based systems) were used for specific applications and these led to the development of "islands" of CAD/CAM (e.g., in drafting, NC, etc.). In 1981, a CAD Pilot Laboratory was set up to gain an in-depth understanding of the emerging concept of integrated CAD/CAM. In the absence of a single-supplier solution, the CAD Pilot Laboratory installed an assortment of selected equipment and tried to integrate the different systems. Translators were built to transfer information between systems. It soon became apparent that this would lead to a far from optimum integrated CAD/CAM system for the product development process. It would also have resulted in the investment of significant resources over many years to maintain and use these systems and their translators through many versions. At about the same time, Xerox carried out a study on requirements for a worldwide corporate CIEMS (Computer Integrated Engineering and Manufacturing System) capable of significantly improving efficiency. It became apparent that the first step to such a system was the implementation of standard hardware and relatively standard software. Turnkey CAD/CAM was seen as a part of the overall CIEMS solution, which would also include engineering-orientated systems for optics, mechanics, electromechanics, electronics, manufacturing, and technical publications. Intergraph was chosen as the turnkey CAD/CAM supplier. It was decided that all design on the CAD/CAM system would be carried out in 3D mode.

It soon became apparent though, that it was not enough to have standard hardware and software, but it was also necessary to standardize the product delivery process. This process was seen to start with the creation of data at the engineering stage, with the need for this data to be shared by the manufacturing, distribution and service, and business functions. Engineers, in turn, required information from the other functions (e.g., to understand the cost impact of a decision to make an engineering change). The CIEMS system, including database and communications software, will become a complete, company-wide information processing, storage, and transmission system.

In 1985, an initial Intergraph system including 20 high-resolution, color graphics workstations, a VAX 11/785, and Versatec plotters was installed at the Rank Xerox site in Welwyn Garden City.

The CAD/CAM facility was centralized, with the workstation user area, CPU room, plot room, and system management and support offices located adjacently. The facility was designed to try to accommodate important human factors. The entire facility was air conditioned to handle the considerable heat load that 20 graphic workstations and the CPU emit. The workstations were arranged in pairs with fabric partitions to prevent reflections between workstations and to allow privacy at each workplace.

Each workplace had a large table (on the left side of 18 work-stations, on the right side of the other two) where drawings/plots could be placed. Two transportable and lockable file cabinets were located at each workplace to allow storage of personal and design reference materials. Each workplace's overhead lighting was operator controlled. Tabletop lighting was also provided. Telephones were provided for each workplace. The windows in the facility were glazed (50% transmission) to reduce heat load and direct and reflected light. Adjustable vertical blinds allowed correction for ambient light. Operator chairs were fully adjustable as were the workstations. Safety, fire protection, and security systems were included. As sophisticated graphics electronics can be quite susceptible to static discharge, static electricity control considerations were taken into account. The facility was run on a 24-hour basis. It has been found to be used most between 8 a.m. and 7 p.m.

The CAD/CAM training program at Welwyn Garden City reflected experience gained during training sessions at other sites. Engineers, designers, and detail draftsmen were all given the same training. There were no discriminatory procedures (such as skill level or experience) for selecting potential CAD/CAM users—the intention being to train as wide a range of users as possible. Each training session started with a 2-day, multimedia CAD/CAM orientation course. This both introduced the trainees to CAD/CAM, and introduced them to the CIEMS environment at Rank Xerox. A few days later, the trainees started a 2-week training course on the Intergraph system. The emphasis of this course was on 3D modelling and the procedures for creating engineering drawings from the 3D models. During this course, the trainees learned how to use the commands available in the system, and how to put the commands together to achieve a required result. As soon as the 2-week training course was finished, the trainees began a 3-month period working with the system on typical problems, but under special conditions. They were not subject to the stresses and strains of working alone under the usual new product development conditions. Instead, they redesigned existing parts from engineering drawings or other data. During this period, experienced users were available to assist, if necessary, the trainees. Once the 3-month period was over, the trainees were expected to use CAD/CAM on a real-life, everyday basis. The complete training session has been repeated with several groups of trainees, and it has been found that more than 90% of the trainees complete it successfully. There is no single criterion of success. In many ways, it is up to the individual to decide if he or she feels more productive with CAD/CAM than without it. Evaluation is further complicated by changes in job content.

A training course has also been implemented for middle management. This consists of a 1.5-day multimedia overview of the CAD/CAM and CIEMS systems, followed by 1.5 days hands-on at the workstation. This has given middle managers (who do not use the system in their work, but will be responsible for products designed on the system) some feel for the capabilities and problems associated with CAD/CAM.

Due to the high proportion of parts manufactured by external suppliers, it has been important to address the problem of transferring data to and from suppliers in the CAD/CAM environment. Data have been transferred directly from the Intergraph system to a supplier's Intergraph system, and via IGES to suppliers using other CAD/CAM systems. Labelled geometry, or APT or COMPACT II formatted data have been transferred to suppliers without CAD/CAM systems but with NC machines. In other cases CAD/CAM system-derived drawings (or plots as they are referred to) have been used to transfer data, but in many cases they have been enhanced by the provision of 3D views and color shading.

Rank Xerox have tried to educate suppliers as to the advantages of using CAD/CAM and have tried to orientate them to using systems with an IGES capability. It is extremely important for the Rank Xerox engineer and the supplier's engineer to work together as soon as possible, and it is felt that CAD/CAM offers a tremendous opportunity to improve the quality of information transfer. However, it is recognized that at present this kind of transfer is complicated, fraught with danger, and that it is difficult to ensure that any given transfer is accurate, complete, and successful in all ways.

It has been found though, that when the engineers can communicate CAD/CAM data, a new problem arises. Staff in "business" functions (e.g., in procurement) have to be educated to carry out their tasks in an environment modified by the use of CAD/CAM.

A variety of special software has been developed to carry out specific tasks associated with design of the particular products. These programs have been interfaced to the Intergraph system, and can be used by the user at the Intergraph workstation.

The calculation of the belt length around a series of pulleys is a common problem in the design of a copier. Manually, this was a time-consuming and error-prone process. A computer graphics program has been developed to carry out the same task quickly and accurately. After the designer has defined the pulley geometry, the program carries out the necessary calculations and the results are displayed. Wrap angle for each pulley, span distance between pulleys, and the total belt pitch length are displayed for the user's information. The designer can try a number of different belt configurations in a short time, thereby obtaining accurate information for the selection of the optimum design.

The timing and synchronization of machine elements are critical to the operation of a copier. Timing diagrams are used to define design requirements and communicate timing information between subsystems. These diagrams, however, become complex and are subject to frequent changes. To reduce the time required to create timing diagrams and to provide the ability to make quick and accurate changes, an automatic timing diagram computer graphics program has been developed. The designer enters signal names and transition times as data, then requests a drawing. The program produces a fully composed and scaled timing diagram showing cycle-up, steady-state, and cycle-down phases. To change the diagram, the designer changes any of the input parameters and requests another drawing.

Early in the design process a space-management layout, known as the "big picture," is created. The big picture comprises the subsystems of the product being designed. The space required by each of these subsystems must be managed within the product's overall envelope, in order to prevent interference and to ensure access for maintenance. This can be done quickly and easily (compared to manual methods) by moving the geometry already created for each of the subsystems into the space-management layout. Volumes required for maintenance access can be defined. The assignment of different colors provides contrast between adjacent subsystems. This comprehensive view of the assembled product helps the engineer check for interferences.

The Intergraph system is a key element in the overall Rank Xerox CIEMS strategy. A significant investment has already been made in purchasing systems and in educating users. This investment is seen to be successful, and plans already exist for increasing the number of systems. Implementation of such a new technology can however only be really successful in the long term, and in the context of an overall company-wide strategy. Different people within the company have different perceptions of where the system is today, and where it should be in the future. Some people may see implementation as being too slow, others (in particular middle management responsible for the development of new products) are not always sure that it is not too fast. Although there may be discussions on the timing of implementation, there is a general consensus as to the overall strategy concerning the world-wide implementation of CIEMS.

8.7 SAAB AIRCRAFT DIVISION

Saab Aircraft, a division of Saab-Scania AB, is based in Linkoping, Sweden and has about 6000 employees. Major products include the Saab SF 340 regional airliner and the military aircraft Viggen and JAS.

Computers have been used by Saab Aircraft for a wide range of applications since the 1950s. In 1985, several hundred computers, including microcomputers, minicomputers, superminicomputers, mainframes, and a supercomputer (a Cray 1) were in use.

Among the first technical applications to make use of computers were NC programming and finite-element analysis. Geometric modelling, electronic CAD/CAM, and mechanical CAD/CAM activities developed relatively independently during their early stages of development. Other major technical computing applications include aerodynamic analysis, flight simulation, and flight testing.

As early as 1955, Saab began using numerical methods for surface definition, and in 1965, a major program was developed in-house for use in defining the geometry of aircraft. In 1970, an electronic CAD/CAM group was set up, and electronic CAD/CAM has been used successfully since then. In 1973 an investigation was carried out by a project group to see how CAD/CAM could be used in mechanical design. As this investigation showed that CAD/CAM could be used productively in the aircraft industry, it was decided to purchase a turnkey CAD/CAM system and use it on the design and development of a new aircraft. Various factors such as problems with the system software and the decision to cancel the development of the new aircraft prevented rapid growth of CAD/CAM. Two workstations were installed in July 1975. Only three more workstations were installed between then and October 1980. (These workstations were however in regular use on applications such as preliminary design of new aircraft, NC programming, and design and nesting of sheet metal parts.) During this period, a decision was taken to replace the Saab-developed Formela surface definition system with BAe's Numerical Master Geometry (NMG).

In 1981, senior management decided that an overall, company-wide strategy for mechanical CAD/CAM should be developed.

External consultants carried out a review of existing operations and proposed that a new approach to CAD/CAM be taken. One of the suggestions was that the system hardware and software be extremely reliable and that data integrity be guaranteed. Another was that a powerful interactive surface modeller be employed. In 1983, a major test of CAD/CAM was undertaken. This proved successful, and in 1984, the decision was taken to go ahead with the installation of CADAM and CATIA on an IBM mainframe. By the end of 1985, Saab Aircraft had, in addition to 18 Gerber workstations, a total of 40 workstations on the IBM. Another 12 workstations were connected to a VAX computer on which the geometry definition program NMG was already installed. At that time, about 15 giga-bytes of disk memory was online to the CAD/CAM computers.

During the period of system selection, it became apparent that there were several ways of justifying the implementation of CAD/CAM. The "strategic justification" ran along the lines that CAD/CAM was essential, therefore the investment was necessary. This reasoning was unacceptable at Saab Aircraft where investments have to be justified on a project by project basis, in this case, on the JAS (multirole fighter) project. It was therefore necessary to justify CAD/CAM with conventional accounting techniques. Although this could have been done by assuming that very high productivity improvements would occur, Saab knew from their previous experience with CAD/CAM that such improvements were both hard to obtain and hard to substantiate. The final justification therefore represented a balanced view of what it was believed would actually result from use of CAD/CAM.

By the end of 1985, the CAD/CAM and Geometry Definition group was made up of about 40 people, of whom 12 were geometry definition specialists. Five people were involved in day-to-day CAD/CAM system operations. A further five people, known as "application specialists," were involved in guiding the introduction of CAD/CAM in new areas, in testing new software releases and in-house developments, in carrying out training, and in helping users to relate system capabilities to application requirements. A total of 12 people were involved in software development. This term covered two major activities—the development of new functions to be included in CAD/CAM systems, and the interfacing of CAD/CAM systems together or to other systems. Due to the high-tech nature of the company's business, many of the functions required were not at that time included in commercially available systems, for example specialized surface geometry functions and functions to design multilayer carbon fiber parts. Although development of such advanced functions was necessary, there were times when it was viewed with regret, in particular when a system supplier brought out a new version incompatible with functions developed by Saab.

The end of 1985 saw CAD/CAM being used on a wide range of applications, including basic aircraft geometry definition, design of complex geometry parts such as air intakes, in providing data for structural and aerodynamic analyses, in the production of wind-tunnel models, for sheet metal flattening and carbon fiber tool design, for nesting, and in preparing technical publications. At that time, CAD/CAM data had been successfully communicated to and from suppliers both on a direct CADAM to CADAM link, and via IGES from ANVIL 4000 to CADAM.

Education and training were important activities of the CAD/CAM group. Some 120 users had received CAD/CAM training by the end of 1985, and 5000 hours of training were scheduled for 1986. Both initial and ongoing training had been given. The amount of initial training given depended very much on the application area and ranged from a few days to a few weeks. An interesting test of efficiency of training was made, revealing the need to link training to practice. Two groups of designers were trained. At the end of the training period, each designer carried out a test example. One group of designers continued to work with the system for one month, the other group did not use the system for a month. At the end of the month, both groups of designers carried out a test example of equivalent difficulty to that carried out at the end of the training period. It was found that designers who had used CAD/CAM throughout the month completed the test four times as quickly as those who had not used CAD/CAM throughout the month. Saab have found that for CAD/CAM training to be really useful and successful, it must be followed by meaningful (but nonessential) work at the CAD/CAM workstation with support readily available from an experienced, "hand-holding" user.

In general, young designers (especially those joining the company direct from college), liked using CAD/CAM, perhaps because it relieved them of an apprenticeship in manual drafting to company standards. It was found that they became productive more quickly using CAD/CAM, than if they used manual methods. More experienced designers, however, were initially reluctant to involve themselves in CAD/CAM with its attendant unknowns resulting from changing technology, task, organization, and status. In many cases though, there was a catalytic effect—once one experienced designer started to use CAD/CAM, others joined in.

A limited amount of ongoing training was given. This training was necessary not only to teach users to use new functions and new modules, but also to ensure that they did not get into a rut of using the system in a less than optimal way.

It has been found that implementation of CAD/CAM involves a wider range of problems than most technical projects. Even when the decision to invest has been taken, a lot of time still has to be spent in convincing people to use CAD/CAM, and a host of organizational and human issues have to be resolved. A major effort has been made to convince middle management that CAD/CAM should be used. These managers, who have committed themselves to producing results by manual methods for a certain date, are often unwilling to take the risk of stepping into the unknown area of CAD/CAM, where they may lack knowledge and control.

Problems have also arisen where tasks have moved across traditional company boundaries. On some occasions, it has been found advantageous for one department to carry out work that would previously have been carried out in another part of the company. The necessary reallocation of funds and resources has had to be resolved much higher up the organizational hierarchy. Geometry definition is now a function reporting to the CAD/CAM Manager rather than to the Design Office. Similarly, the NC programming support function which reported previously through the Production Department now reports to the CAD/CAM Manager. In some cases, the power of the CAD/CAM system has made it possible for designers (particularly in tool design) to develop NC tool paths themselves without assistance from NC part programmers.

Difficulties have been encountered owing to the lack of standardization, not only among CAD/CAM systems, but also between CAD/CAM systems and existing manual systems. For example, the existing company part-numbering system, using a maximum of 22 characters, was modified as CADAM could only handle a maximum of 20 characters. Similarly, because the CAD/CAM system did not support some internal company standards, the Standards Department redefined some standards or accepted departures.

The need to transfer data from one CAD/CAM system to another has led to special interfaces being built between the systems. Even more importantly, the use of really powerful data management and control systems has been impeded by the lack of compatibility of data structures in the different systems.

System monitoring has been carried out with the intention of improving performance, but it has been found difficult to identify parameters that are really useful in evaluating the results of system usage.

The continuing rapid change in CAD/CAM technologies, whether it be in the areas of geometric modelling, software, computers, or workstations has limited the extent to which forward planning of CAD/CAM has been possible, but a detailed two year CAD/CAM development plan has been maintained.

One of the major gains resulting from the use of CAD/CAM is that tasks that could not be carried out before (or only with great difficulty) (e.g., the design of carbon fiber parts and tools), can now be approached with more confidence. Complicated geometries, such as those of air intakes, can now be designed more easily and analyzed in greater detail. Interference studies of moving parts, such as the undercarriage, have been carried out in more detail. Savings have been achieved just because it has been possible to display a 3-D wireframe representation of a part on the screen, thus making early identification of design errors possible. Another major source of gain has come from the increased speed with which modifications have been made to existing designs.

Gains have occurred when previously separate steps in the overall design and manufacturing process have been integrated. One example of this is the way in which the same geometry has been used in aerodynamic analysis and to NC machine wind-tunnel models. Another example is the use of the same data to design airframe parts, to design the tools to bend sheet metal to produce these parts, to nest the parts, and to produce NC data to cut these parts out of sheet material. A similar chain has been developed for carbon fiber parts.

Investments in CAD/CAM will continue to be made by Saab Aircraft. Future developments planned include increasing the number of users of CAD/CAM, increasing the number of applications handled by CAD/CAM, and strengthening the links between islands of CAD/CAM-assisted areas in the overall design and manufacturing engineering process.

8.8 SALOMON SA

Salomon SA is a world leader in winter sports equipment. Based in Annecy, France the company had about 1400 employees in 1984 and a turnover of about 1,650 million French francs. Major product lines were downhill ski boots and bindings, cross-country ski boots and bindings, and golf clubs. The company attaches great importance to innovation, quality, and excellence. As a result, the company has more than 130 design and manufacturing engineers. The most modern design and production techniques are used.

In 1981, a team of 5 people was organized to identify the CAD/CAM needs of the different product divisions. At that time, it was decided that CAD/CAM would not be handled by the EDP Department but by the Technical Services Support Group. Since 1981, between 2 and 4 support staff have assisted users.

The two major objectives of CAD/CAM were to reduce the time taken to get a product from initial design through to production, and to improve the quality of the product. (With manual methods, some 70 man-months elapsed between the production of an initial clay model of a ski boot and the corresponding range of moulds for production.) It was always the intention to use CAD/CAM to go directly from design to machining without the production of intermediate drawings.

A 2-workstation Computervision CADDS4 system was installed in early 1984 in the ski boot Design Office. Initially, two designers and two Technical Services Support staff were trained to use the system. These four people then spent some four months building up a support structure so that the system could be used efficiently in the Salomon environment. Utility programs were developed, naming conventions introduced, menus modified, plotter routines implemented, and so on.

From June 1984 onward, 12 designers followed training courses in Computervision's Paris office. Their major requirements of the system were the complete definition of ski boot geometry, the possibility to carry out detailed exploration of alternative designs, ensuring that there was no interference between moving parts, automatically developing all sizes of a boot model from the basic shape designed for one size, simulation of plastic flow in a mould, and the calculation of the resistance of particular parts of the boot to given stresses. Color shading functions were needed to check the continuity of adjacent complex surfaces.

To prove that the system was really capable of meeting these requirements, data from the first boot designed was used to design (with the system) the corresponding moulds. Tool paths were then calculated on the system. These were transferred via a modem to a DEC-based machine tool controller. The moulds were NC machined. The total time from design to machining was four times less than with traditional methods.

Once it had been shown that the system could be used by all the designers, and that it could be used from design through to machining, 2 more workstations were purchased. Mould designers and NC machine tool operators were then trained to use the system. This time the training was given internally by Salomon's own staff. Generally, four people were trained at a time. Each user followed a four-day basic training course and further courses as required (e.g., 3-day complex surface course, 3-day machining course, 3-day dimensioning course to learn how to prepare drawings for subcontractors). It was found that after the initial training period new users required assistance for several months, and it was only after about 6 months that they could really use the system without any assistance. After about another 6 months they became expert users. Apart from courses given to users, about 40 people in the company attended CAD/CAM introductory courses. The result of this investment in training and education is that everyone in the company supports the use of CAD/CAM, and that all users are willing to use the system. The system is currently available for about 12 hours per day, though individual users rarely work more than 4 hours per day at the screen.

Once the mould designers had learned to use the system, it was found that the amount of misinterpretation of geometry decreased greatly and that by building up and reusing a library of standard parts, even more time could be saved.

The initial decision to invest in CAD/CAM was a strategic one and a very detailed cost justification of CAD/CAM was not carried out. Looking back, it is believed that this was the right approach. It saved time and money that would have been spent justifying system purchase. During that time, a tremendous amount of experience was gained with the system. Many very productive applications of the system that were not initially identified have now been developed.

Systems use was sufficiently successful for a fifth workstation to be purchased in October 1985 along with 2 IBM PC AT computers for MICROCADDS. The PCs will be used in early stages of conceptual design, in training, and in hands-on learning.

At the end of 1985, the intention was to expand the 5 workstations, APU, CPU, 2 x 300 MB disk CADDS 4 installation so that staff from other product divisions could start to use CAD/CAM.

8.9 STELLRAM SA

Stellram is a Swiss company with about 280 employees at its major sites in Nyon and Gland. The principal activity of Stellram is to produce cutting tools with indexable tungsten carbide inserts. The company first used computers (2 HP21 MX) for business applications in 1976. These computers were later replaced by an HP3000.

In early 1980 it became apparent that new technologies were needed to generate the data for NC machining of indexable carbide inserts tooling. A carbide insert is the part of the tool that actually cuts the metal. Its relatively complex shape, known as the "cutting geometry," defines the shape and length of chips. During the manufacturing process for the carbide inserts, which involves EDM, an electrode corresponding to the carbide insert is milled on a 3-axis NC machine. The positioning of some carbide inserts on specific tools can be quite complex. In fact, in cases where a succession of inserts follow each other on a simultaneously tapered and helical path, the 3-D location of each individual insert pocket can be extremely difficult to draw up and to locate in space.

When Stellram looked for a CAD/CAM system in 1980, major requirements were for a system that could be used for the design of complex 3-D surfaces and the production of the corresponding NC tapes. The systems UNISURF (for design) and SURFAPT (for machining) were selected. They were run off a graphic screen in the Nyon offices of Stellram, connected via a modem to CISI computers in Saclay, near Paris, France. With the aid of these programs, a new series of cutting geometries was designed and produced. However, the poor quality of overall data transmission (from Saclay via Transpac, and then via Euronet to Zurich, and then by telephone line to Nyon) gave rise to some problems, and after one year's use of the systems, it was decided to look for another solution.

In 1982 the major requirements were for a system that could be run in-house, if possible on an HP computer, and that could design complex surfaces and generate the corresponding tapes for CNC 3-axis machines. Other requirements were that the system be highly interactive and easy to use, capable of producing drawings and basic 3-D design of noncomplex shapes. In addition, the system would have to be able to generate, via a special postprocessor, programs for a 5-axis CNC machining center.

A one-terminal IDS 80 system from Gerber (with an A1 plotter) was selected and installed in the Gland offices in late 1982. Four people were given four days initial training, and a further two days training one month later. After the first training course it was found to be possible to use the system quite effectively in everyday work. The few problems that arose during the first few week's work were resolved during the second course. Two design engineers and a NC programmer were among the people trained. The fourth person was the Carbide Insert Product Manager, who was appointed CAD/CAM manager. After the system had been used successfully for some time, a second terminal was installed, and a design draftsman trained.

The CAD/CAM system has been applied to design many new tools. The system allows tools to be designed more productively, and allows the creation of designs that would be virtually impossible by manual techniques. Once a design engineer has built up a three-dimensional model of a tool, this data is passed directly to the NC programmer and the design draftsman. Thus, the same engineering model is used in design and manufacturing. This eliminates problems due to mistakes in producing and interpreting drawings. The NC programmer uses the model to produce the tool path, which can then be simulated on the graphics screen. It is then passed through a postprocessor to generate the ISO NC program. The three-dimensional model is used by the design draftsman in the production of manufacturing drawings.

One of the major uses of the system is to design the cutting geometry of carbide inserts. Input data for a new geometry comes in the form of a set of parameters (angles, widths, coordinates of points on sections, etc.) from the Testing Laboratory. The Gerber system is used to build up sections and to design patches of surface (a typical geometry contains 50 to 100 patches). These patches are used to generate the path of a mill used to machine the corresponding electrode.

Stellram has written a specialized CAD/CAM program (in FORTRAN) to run on the HP 3000 computer and a Tektronix graphics screen. From the set of parameters received from the laboratory it automatically calculates the NC tool path, and displays it on the screen.

The IDS80 version used by Stellram treats surfaces to be machined on a patch-by-patch basis, and does not have a circular interpolation option. As many of the surfaces are cylindrical in nature, Stellram has developed a special postprocessor which analyzes the tool path points output from IDS80 and, where possible, replaces them with the parameters of the corresponding arc. This results in a shorter tape being produced, and in less wear on the machine.

Stellram has found CAD/CAM to be of great benefit and would like to extend its use to calculate the shape of chip that will be produced by a given tool design under certain cutting conditions. At the end of 1985, this feature did not exist, thus prototype tools still had to be built. This slowed down the rate at which even better cutting geometries and tools could be developed.

8.10 A SWISS WATCH MANUFACTURER

This company is involved in the design and assembly of watches, and manufactures the watch movements and cases in its factories in Switzerland, whereas the production of other parts is subcontracted to a large number of fairly small, traditional, watch-part suppliers. Not many of these companies are equipped with CAD/CAM systems, and they need to receive information on traditional manufacturing drawings.

The advantages of some form of CAD/CAM having been identified for this Design Office, several CAD/CAM systems were investigated during 1983. They were rejected, mainly on the grounds of being unnecessarily complicated for drawing office work. Then a demonstration of TELL (a system developed by Polysoft Consulting) was attended. TELL is basically a 2-D drafting system. It corresponded well to the company's requirements. For example, it is easy for draftsmen to learn and use, it can be used to produce high precision drawings for subcontractors, and it does not require specialized EDP support staff. It offers facilities for repetitive design (e.g., for the cogs of a wheel), reuse of standard parts (e.g., screws), and user modification of menus. In addition, as the system runs on standard HP computers and had a relatively low price, purchase did not involve a major risk.

The system was installed in a one workstation, diskette-based configuration in January 1984. It was installed in one morning, and in the afternoon, one user (a draftsman), was trained and introduced to the system documentation. Two weeks later, the user could develop acceptable, production-standard drawings with the system as quickly as by hand, and four weeks later had completely mastered the system. In August 1984, a second workstation was installed. The two workstations were connected via a network to a 132 MB disk unit, an A0 plotter, and an A3 plotter. (Although most drawings only required about 20 KB to 50 KB of storage, some drawings had exceeded the capacity of the diskette, and it was decided preferable to invest in a larger data storage unit).

Design Office staff have built up a library of standard parts (e.g., screws and standard drawing symbols) and have configured the tablet menus so as to have rapid access to the most often used standard parts and functions. They have also defined the organization of drawing files on the disk so that drawings can be accessed as efficiently as possible.

The system is used very much as a tool for producing drawings faster. The drawing numbering system used is the same as that for manually produced drawings. Users enter the drawing number at the keyboard and it is stored along with the drawing. In the two years since the system was introduced, about 3000 drawings have been produced. They take up about 50 MB of disk space. The disk is backed up onto magnetic tape every week.

Experience shows that it is no quicker for a draftsman to draw a drawing with the system than it is to do it manually, unless some degree of repetition is involved. The major types of drawings produced with the system are detail part drawings, assembly drawings (including up to 150 parts), and control drawings. Drawings are also made for kinematic analysis studies of moving parts and for a line-following prototype production machine. Major benefits from use of the system occur in the use of standard parts on detail drawings, in repetitive design (e.g., of cogs on a wheel, or indices on a face), in parameterized design (e.g., of cogs), in the production and modification of assembly drawings, and in the production of large-scale (20 to 1 or 50 to 1) control drawings where it may be necessary to draw arcs of radius 2 m to a precision of 0.05 mm. As an example of the time savings achieved, it has been noted that assembly drawings that took 5 days to produce manually, only require about 3 hours with the system.

One problem that has been encountered is the difficulty of interfacing hardware from other computer equipment manufacturers to the system. Another problem concerns the pen plotter. Sometimes the nibs scrape off a thin layer of paper and then do not draw as clearly as they should. On other occasions, if care is not taken, the ink reservoirs run dry during a plot.

By the end of 1985, the majority of designers and draftsmen could use the system. The system was used from seven o'clock in the morning until seven o'clock in the evening. The major users of the system tended to be the draftsmen. They often spend eight hours a day at the workstation. In the future, the system will be expanded, and more designers and draftsmen will be trained to use it.

8.11 WESTLAND HELICOPTERS LIMITED

The headquarters of Westland Helicopters is in Yeovil, Somerset, England. About 7000 people are employed in the marketing, design, manufacturing and sale of helicopters such as the Sea King, the Lynx, the WG30 and the EH101.

The first scientific computing programs were developed in 1962. Since then a wide range of technical computing programs embracing many complex disciplines has been developed. Some are very large number-crunching programs for aerodynamic and structural analysis, others just small interactive packages used by dynamicists to obtain very quick solutions to one-off problems. Programming of NC machine tools started in 1965.

In 1975 a committee was set up to investigate the possibilities of using CAD/CAM. Some of the major systems then available were tested, but they were found to be very expensive, and their purchase could not be cost justified. However, a small CAD/CAM system team was set up and a CAD system for electrical drafting was designed and commissioned. Trade-marked 'Electricad' it is still in use. A CAD/CAM software package was purchased and installed on the same 16-bit computer as 'Electricad.' Apart from leading to productivity gains, development and use of these systems was beneficial to the company in that it increased the general level of CAD/CAM knowledge and understanding. However, CAD/CAM did not have a tremendous impact, and by the end of 1979 only 6 workstations were installed and in use.

In 1980, a committee comprising senior executives of the company was formed to develop an overall strategy for CAD/CAM implementation and use. The strategy proposed was approved by the Board of Directors, and the necessary funds for a major implementation of CAD/CAM were made available. As a result, during the period 1980–1985, several major CAD/CAM systems were purchased and implemented by a team with the necessary expertise to integrate these systems.

In 1980, it was estimated that there were about 200 potential users of CAD/CAM in the company, and that they would require at least 40 workstations. The use of CAD/CAM was so successful that rapid growth occurred in the number of users, the number of workstations and the number of applications handled with CAD/CAM. By January 1984, 72 workstations were in use, and it was estimated that there were 400 potential users of CAD/CAM. At that time, more than 9000 hours of CAD/CAM training had already been given at the screen. At the end of 1985, more than 80 workstations were in use, and some 500 people had been trained.

Excluding local hardcopy devices, the plotters in use were 2 Benson drum plotters, 1 Calcomp drum plotter, 2 Versatec electrostatic A0 plotters and a Kongsberg 72 inch × 98 inch flat bed plotter. Some 15,000 plots per month were produced on these plotters in 1985.

Apart from lofting of aircraft lines and structural details, CAD/CAM has been used for applications such as design of gear boxes and transmissions, flying controls, structures and undercarriage mechanisms. CAD/CAM is used to produce drawing-based information associated with aircraft wiring diagrams. As a component may exist on several drawings, an Electricad database has been built up for each aircraft with a one-on-one relationship between components and computer records. From this database it is possible to extract component and cable information, as well as to detect interdrawing mismatch of information. CAD/CAM techniques are also used in automated loom manufacture.

In the manufacturing engineering area, CAD/CAM has been used for applications such as the development of machine fixtures, assembly jigs and checking gauges. It is used in programming NC machines (e.g., for milling blade moulds), programming composite tape laying machines, nesting sheet metal parts, and for some factory layouts.

Activities in manufacturing engineering have been considerably modified and extended as a result of the availability of design information in digital form. NC programming, previously carried out using batch systems, can now take place interactively at graphics workstations. The Cutter Location files are then post-processed to provide numerical control programs. DNC systems have been implemented in which data held on a central computer is passed, via a mini computer, directly via fiber optic lines to the machine tool controller.

In the Detail Manufacturing Complex, these developments have been taken one step further. In this Complex, a CAD/CAM program is used to define and manage geometry for 2-D sheet metal components, and to generate the corresponding Cutter Location files. These files are then passed to a computer-aided nesting program, which nests the individual components on large sheets of material for subsequent machining on NC routers. The integration of this activity with computer-based production control has led to a fully computer-controlled manufacturing facility.

The CAD/CAM Manager, reporting to the Engineering Systems Executive, leads a 15-strong team giving full technical support to users. The CAD/CAM team members are engineers as well as CAD/CAM experts and most of them have experience of design and drafting. It is proposed to develop CAD/CAM within a Host Structure concept. The philosophy behind this concept includes the need to insulate the Company Data Base from the CAD/CAM systems being used to generate the geometries stored in it, and the provision of a coherent framework within which the various CAD/CAM systems may intercommunicate.

In general, graphics workstations are used to produce graphics, and alphanumeric workstations for the input of alphanumeric information. Thus parts lists for assembly drawings are entered at alphanumeric screens to generate an "as drawn" database, while the assembly drawings are produced on graphic screens. This database is used to produce schedules and kits of parts for the Production Engineering Department.

Although file-based CAD/CAM can be used productively it is recognized that a relational engineering database will be required in the future. This database will include technical and performance information, product geometry information, textual product information and general management information. The engineering database is a major development area for the CAD/CAM team.

Experience has emphasized the greater need for discipline when using CAD/CAM than when using traditional methods. Success has been found to be very dependent on people and applications, with the greatest success coming in manufacturing engineering areas. In these areas there is always willingness to use CAD/CAM as a tool to carry out the job.

It is expected that at least 700 people will be using the system in the near future, and this requires a major effort in getting people's attitudes to change. It is found that not only do some users not want to use CAD/CAM, and middle management not want to risk its causing them to miss targets, but industrial relations pressures and empire-building managers impede the structural changes necessary for total success.

In the future Westland will continue to develop CAD/CAM, but extra investment will be made in using CAD/CAM at early stages of development and in ensuring that such use really is as productive as possible.

Westland has taken major steps along the road to CAD/CAM, in particular since 1980, but recognize that in today's quickly changing environment, further development of CAD/CAM represents a necessary and major challenge, not only to management but to the whole company.

8.12 A WHITE GOODS MANUFACTURER

The company makes consumer-oriented, high technology products and obviously needs to take full advantage of computer-aided techniques such as CAD/CAM to be competitive in international markets.

Although I had an appointment to see the Managing Director I was not able to meet him. Like the Marketing Director and the Sales Director he was out making sales calls. The Finance Director could not see me as an urgent meeting had just been called by the company auditors. I was met by the Personnel Manager, a charming lady who gave me an overview of the company that she must have given to many potential employees. She was sorry that I would not be able to meet any of the directors, but was sure that the DP Manager would be able to give me all the information I needed. She took me over to another building where a harrassed-looking lady, who I learned afterwards was the DP Manager's secretary, asked me to wait in a corridor while the DP Manager finished an urgent memo to his boss, the Finance Director. After telling me what an important man the DP Manager was, she excused herself saying that she had to hurriedly retype a summary of the last six months' usage of the computers.

Eventually, the DP Manager invited me into his office. I explained my visit and he said that he was sure that he could help me a lot, after all it was due to him that the company used CAD/CAM. I sat down on rather a dirty looking chair, and as he talked, I looked round his office. Old listings filled the shelves and empty boxes littered the floor. He noticed my regard settle on a three-foot long turbine blade propped up in a corner. "That was a benchmark that we did for the CAD/CAM system" he said. "I didn't know this company made turbine blades," I said. "We don't," he replied, "but we really wanted to test out the system, and if it can handle a turbine blade it can handle our parts."

I would have liked to explain to him that there was no sense behind what he had just said, but he had already changed the subject.

"You know," he was saying "the company is really very lucky to have CAD/CAM." It turned out that a few years back, he had ordered a new computer for business applications, but the order had fallen through because the manufacturer had pulled out of the market. Rather than waste the money, the DP Manager told me that he had decided to buy a very good two workstation CAD/CAM system. The software was developed by a company that had only just been set up, and he had managed to buy the system for a relatively low price as he knew that the company needed some reference sites. He thought that the system was very good. There was always someone using it, and it produced lots of drawings. He stood on a chair to look through some rolls of drawings on the top of a cupboard but apparently could not find the one he wanted.

I asked about initial use of the system, and he told me that there had been few problems apart from the lack of documentation for the system. However, he felt that this had stimulated the users to understand the system in depth, and after all, as he said, people are paid to use the system not to read the documentation. Drawing management had not been a problem. Drawings were stored under the file management system of the computer. As far as he knew all users looked after their own files. There were about 15 users of the system. That was a very small number compared to the business side of DP where there were 58 terminals attached to the business computer, and 97 users. I asked about NC, but he told me that the part programmers had found that the CAD/CAM system could not handle some of their parts, and had bought themselves a separate microcomputer-based system. That was just as well, he said, since another 6 users on the CAD/CAM system might have caused some problems.

"One big problem I have now," he said "is that the DP budget has been cut back, so it is not possible to expand the use of CAD/CAM." He told me that it was not possible to buy any more workstations for the existing computer, so that they would have to buy a bigger computer if they wanted to expand the system. As this computer had not been budgeted for, the Finance Director would not understand why it was needed. He went on to tell me about another problem, the lack of good operators for CAD/CAM systems. They had had three or four CAD/CAM operators who had left as soon as they began to be really useful, and he did not understand why.

I told the DP Manager that I would like to have a look at the CAD/CAM installation. "Of course," he replied, "but have you any other questions I could answer." I asked him about the way in which the system was used, and if the users' experiences with the system were used to improve performance. He told me that the system was just used during normal working hours as he could not afford to pay for another system operator. Once, he had been to a user group meeting and talked about use of the system, but no one had been interested so he had not gone again. If the users had any complaints about the system he was always ready to listen to them, but he thought the last complaint had been made about a year ago, and he could not remember what it was.

He thought that CAD/CAM had been a real success—people from local companies interested in CAD/CAM often came to look at the system. It really was very good, he would get someone to show me how it worked. As he led me down to the computer room, I asked what the CAD/CAM strategy for the future was. He replied that he hoped that when more money was available it would be possible to buy another computer and some more workstations.

We entered the computer room where two operators were drinking coffee and smoking. "Not finished your break, yet, boys," shouted the DP Manager. "No," one of them shouted back, "we started late because we had to fix that disk unit. It's causing problems again." I asked the DP Manager why the room was so noisy, and he told me that some ventilation fans seemed to be broken, but he had no money to pay for them to be repaired. "We don't have a maintenance contract," he said. "Arm and a leg—and you get nothing for it. Come on, let me show you the workstations."

It had been decided to site the workstations near the plotter for convenience, but due to lack of space, it had been necessary to put them in a corridor. When we arrived both workstations were in use. In front of one of them a man introduced to me as the Drawing Office Manager was in heated discussion with the user. It was difficult to see what they were looking at on the screen as the image was very faint. It turned out that the user could not recall anything from a previously stored library of parts. "Are you sure your library of parts is on the disk," said the DP Manager, "you know the operators have some trouble with the disks." Yes," replied the user, "I've just checked and the file is there." The DP Manager suggested that they call another company which had the same system to try to find out what the problem was. However, the user had already done that and been told that there was no problem with the software; it was the hardware that was at fault. Apparently the parts were being drawn on the screen, but with the beam off, and so they could not be seen. It seemed to the user that the system vendor should be called. The DP Manager did not want to call the vendor (probably because he had not paid for a maintenance contract). Instead, he went off to find one of the operators who knew something about hardware and might be able to put things right.

When he had gone, the Drawing Office Manager said, "It's always the same with this system. If we'd done this job by hand, we would have finished it last week. These computer guys are useless, they just don't understand what we're doing. This is the last time we use this system." The user threw the stylus down on the tablet: "That's another week's work wasted. Why can't we get a decent system." The other user, a rather small lady working at the other workstation suddenly said "Hey, you guys come and look at this."

She looked very uncomfortable, sitting on a wooden stool, and I am sure that she found it difficult to access all parts of the tablet. We moved over to see her screen better. "No," she said, "that's not what I wanted to show you. What's happening now? Is this a bug or just poor response time?" Suddenly a picture flashed up on the screen. "Look," she said, "what's this rubbish I found in the data base. Someone's designed a part that's much too thin. That would break after a few days' use. I'm going to change it. Let's double the thickness." "Hey, leave that alone," said the other user, "that's mine. I spent all last week doing that. Where did you get it from." "I don't know," she said, "it just appeared when I picked this menu item." As she pointed to the tablet, the screen went blank and a few seconds later, a message came up on the screen: "X023-Routine SPALT Pass 3 Error 6." As the three of them started a lively discussion of what to do next, I slipped away. I felt that I had learned enough for one day about how not to use CAD/CAM.

9

INGREDIENTS FOR SUCCESSFUL IMPLEMENTATION OF CAD/CAM

This chapter is based on experience with companies of different sizes, in different engineering sectors, and in different countries. In addition, these companies have been using CAD/CAM for differing lengths of time. They therefore represent a good cross-section of mechanical engineering CAD/CAM system users.

The major ingredients for successful implementation of CAD/CAM are described under the following headings:

Top management
Basic applications knowledge
Requirements analysis

197

Table 9.1 Ingredients for a Successful CAD/CAM Implementation

9.11 Ongoing management			
9.7 CAD/CAM system	9.8 CAD/CAM system support	9.9 CAD/CAM system development	9.10 CAD/CAM system operations
9.4 Users	9.5 Focused applications	9.6 Training	
9.2 Basic applications knowledge			
9.3 Requirements analysis			
9.1 Top management			

Users
Focused applications
Training and education
The CAD/CAM system
CAD/CAM system support
CAD/CAM system development
CAD/CAM system operations
Ongoing CAD/CAM management

In Table 9.1, these ingredients are shown as building blocks forming a wall. If the blocks at the base of the wall are not sufficiently developed, then the blocks above them will not be stable. As the use of CAD/CAM within a company goes through different stages of evolution, the relative importance of the ingredients changes.

9.1 TOP MANAGEMENT

It is good management at the highest level that has the most effect on the implementation of CAD/CAM.

9.1.1 Commitment to CAD/CAM

Top management must be committed to the introduction and use of CAD/CAM, and must make this commitment highly visible. It is not easy to successfully implement CAD/CAM in a company, and in the absence of a clear lead from the top, there will be many middle managers and potential users who will be only too happy to scupper CAD/CAM. Important demonstrations of commitment by top management include giving full support to the CAD/CAM manager, and requiring middle management to use CAD/CAM in their departments. Management must not only give a lead to the rest of the company but also ensure that the necessary decisions and actions are taken. This is necessary at the time of system selection, since many companies have a tendency to refine their investigations unnecessarily rather than actually get on with the implementation. It is equally necessary after system installation, at which time many companies believe that just because the system is installed, it is working optimally and needs no further attention. This is generally not so, and top management must ensure on an ongoing basis that productivity gains really are being achieved. In the absence of such gains, corrective action must be taken.

9.1.2   A Global Approach to CAD/CAM

For CAD/CAM to succeed, computer-based product information must be able to transit between different stages of the design engineering and manufacturing engineering process. However, in many cases, CAD/CAM has only been implemented in just one, or in a limited number of these stages (e.g., in design engineering but not in manufacturing engineering), or in the "proposals" drawing office but not in the "production" drawing office. Partial implementation of this sort can only be successful on a very limited scale. To avoid getting into situations of this kind, top management which, alone, has overall responsibility for different stages of the process, must lead the company to a more global approach to CAD/CAM.

9.1.3   A Long-Term CAD/CAM Plan

Without a well-defined CAD/CAM plan, there is little chance of a company meeting its CAD/CAM objectives. The responsibility for such a plan lies with top management. Although 20 years ago, definition of a CAD/CAM plan may have been extremely difficult, the weight of CAD/CAM experience now makes it possible.

9.1.4   Correct Financial Understanding of CAD/CAM

Once the decision has been taken to implement CAD/CAM, it is important to ensure that CAD/CAM services are not charged to users at a prohibitive rate. Overcharging may result in a financial "profit" but can discourage use of the system, and therefore make a nonsense of the initial objective to increase overall productivity.

During the system selection phase, when an economic benchmark is carried out for each system, top management must make it clear to all concerned that indirect productivity gains should be included in the calculations.

9.1.5 The Management of Change

The successful implementation of CAD/CAM requires changes in the organizational structure of the company, and the way in which certain tasks are carried out. New types of staff must be employed. The relationship between design engineering and manufacturing engineering will change. However, most people in the company are quite happy with the status quo, and although they enjoy complaining and criticizing, do not really want changes to be made. After all, changes can lead to unpleasantness such as demotion, more work, and unwanted responsibility.

Only top management has the motivation and the power to effect changes. Whereas many of the factors concerned with successful CAD/CAM involve management lead and follow-up with others carrying out the actions, top management itself must act to carry out the necessary changes.

9.2 BASIC APPLICATIONS KNOWLEDGE

A CAD/CAM system is just a tool that allows users to carry out their everyday tasks more productively. It is not a piece of magic that suddenly makes untrained personnel become highly skilled design and manufacturing engineers. If personnel do not have a good basic knowledge of how to carry out their tasks without CAD/CAM, then even the best system cannot be expected to improve their performance.

A company should not expect that implementation of CAD/CAM will lead to a reduction in the number of highly qualified staff, and their replacement by less qualified personnel. If anything, the opposite is true, with the system providing the most important productivity benefits when used by highly qualified staff.

Similarly if the company does not have competent staff capable of managing projects, use of the system should not be expected to automatically result in better project management.

9.3 REQUIREMENTS ANALYSIS

In view of the global, company-wide nature of CAD/CAM, it is very important that before choosing a CAD/CAM system, a company fully understands its real CAD/CAM needs, and where and how CAD/CAM should be used. Unfortunately, experience shows that a company's knowledge of itself is very fragmented, with some individuals having deep knowledge of the particular area in which they work, but virtually no one having a detailed overall view of the company. The information required must be built up by the task force. It cannot be obtained directly from external sources. System vendors can supply part of the answers, but even this may be somewhat partial. One solution lies in using experienced consultants to assist the company to know itself better.

Unfortunately, many companies do not carry out the analysis of requirements in sufficient depth. This may result in selection of an unsuitable system. In other cases it leads to the formation of opposing teams within the company, each team believing its (generally biased) interpretation of the requirements to be the only correct one. Often an attempt is then made within the company to defuse the situation by carrying out a further analysis of requirements. To avoid such problems, the initial analysis should be carried out in sufficient detail and by a mixed team of people representing all disciplines concerned.

Once the requirements analysis has been carried out it should be used as a basis for making a decision, not as a basis for endless discussions and meetings concerning its interpretation.

9.4 USERS

The users of the CAD/CAM system have an important role to play. If they do not know how to use the system efficiently, or for some reason do not want to work with it, then productivity will suffer.

9.4.1 User Awareness

Many people in the company are opposed to change of any kind, and so will automatically be opposed to the introduction of CAD/CAM. This attitude must be overcome since it is through the users of the system that productivity gains will be made. Potential system users should be informed as early as possible that the introduction of CAD/CAM is under consideration. They should be told why it is needed and what effects it should have on the company.

9.4.2 User Motivation

Having made the users aware of the intention to investigate the use of CAD/CAM, the next step is to motivate them to use it as efficiently as possible. They should be assured that introduction of CAD/CAM will not result in layoffs. If the users believe that CAD/CAM threatens their livelihoods they may refuse to work with it or use it very inefficiently to ensure that it does not result in high productivity gains. Incentives for highly successful users should be considered, although care must be taken not to penalize less successful users.

9.4.3 Superusers

Some users invariably understand the system better than others, and enjoy working with it. With a little extra training, these super-users can become extremely effective both when addressing project work and when assisting other users to be more productive.

9.4.4 Helping and Listening to the Users

Learning to use a CAD/CAM system is not an easy process for the majority of people. It is necessary to help them as much as possible especially in the early stages when their self-confidence may be at a low level. Once the system is installed and running, it is important that information on everyday performance and use be collected, analyzed, and then acted upon. One way of doing this is through regular meetings of users, project managers, and the CAD/CAM manager. Problems with the system must be identified and solved as quickly as possible.

9.5 FOCUSED APPLICATIONS

It is not possible initially to simultaneously apply CAD/CAM techniques to all operations in the company. Since it is necessary to choose where to start applying CAD/CAM, it is best to apply it where it will be most successful. Within each company, there are applications that can benefit from CAD/CAM, and applications that will be less productive with CAD/CAM than without it. All applications liable to be impacted by CAD/CAM must be identified during the initial investigation into the use of CAD/CAM. An overall plan should be drawn up detailing the order in which applications will be transferred to CAD/CAM. Focusing first on applications that will benefit most from CAD/CAM leads to productivity gains in these areas, helps build up a successful image for CAD/CAM, and causes the CAD/CAM experience and knowledge levels within the company to grow. This CAD/CAM experience and knowledge, and the good image of CAD/CAM within the company, will be of great assistance when it comes to applying CAD/CAM to areas where immediate productivity gains are less obvious.

It must be stressed however, that focusing on particular applications should only be carried out within a well-defined overall plan. Random focusing, and focusing on particular applications without due regard to requirements of other applications can be disastrous.

9.6 TRAINING AND EDUCATION

Training is one of the most neglected areas in implementation and use of CAD/CAM. One reason may be the relative ease with which a training budget can be slashed and consequent "savings" produced. Nearly all advanced manufacturing technologies require a high level of training since not only are the technologies complex in themselves, but their users have had virtually no formal education related to them. Cutting back on training can invariably be equated to cutting back on productivity increases. It is very difficult to learn about a new technology entirely on-the-job, since what is needed is as much an understanding of the basic processes involved, as just operating knowledge.

9.6.1 Company-wide Education and Training

It must be remembered that initially no one in the company will know anything at all about CAD/CAM. Users have to learn to use it and middle management must learn how to apply it to projects. The CAD/CAM manager and the CAD/CAM team need to learn a wide variety of new skills. Top management must learn enough to understand what can realistically be done with CAD/CAM within the company, what can be expected from application of different levels of resources, and how to manage the effects its introduction will have on the overall operations of the company.

9.6.2 Ongoing Training and Education

Training and education are necessary during the selection process, before the system arrives, during the period immediately after installation, and then as long as the use of CAD/CAM techniques within the company is expanded and new users are introduced to the system.

Ongoing training is necessary partly because improved versions of the system will be installed and must be understood, and partly because after using the system for a certain time, users are capable of using it in a different, more productive way. People throughout the company must learn how to handle the way that other systems in the company are interfaced to the CAD/CAM system. In addition, training is necessary to be able to further extend the use of CAD/CAM throughout the company.

9.7 THE CAD/CAM SYSTEM

It is not always necessary to use the *best* available CAD/CAM system. More importance should be attached to selecting and using the system that fits the company's requirements. Whenever possible, user companies should avoid developing their own systems. Most of the CAD/CAM functions required by the average small- to medium-sized company are available in systems that can be purchased. The effort required to develop and maintain a CAD/CAM system is enormous, and user companies should concentrate their resources on using CAD/CAM to improve their products, and not on developing CAD/CAM software.

The major characteristics of the system that will affect its use and performance are as follows.

9.7.1 System Hardware and Software

The most important requirements of the system are that it offers the functions required by the company, is reliable, and can be used by people within the company. The system should make use of relatively modern hardware and software technologies. If it uses out-of-date technology, one can expect the vendor to make major changes (probably disruptive) in the near future.

9.7.2 Geometric Modelling

The geometric modeller used within the system should be capable of modelling the type of products that the company produces. There is often a conflict of interests when selecting a modeller. On the one hand, many people used to working with paper as a primary medium for geometry definition will want to continue using similar techniques (e.g., a 2D system). On the other hand, most products of companies involved in mechanical engineering are three-dimensional, and, in the future 3D modellers will be favored as they can handle more complete product information. A similar problem arises if, as is often the case, the company makes a variety of products that have different needs from the point of view of geometric modelling. For example, a plastic injection mould may need a surface modeller, whereas wiring drawings could be produced effectively with a simple 2D modeller. The ideal solution is to use a system containing a range of compatible modellers meeting company requirements.

9.7.3 A Potential for Interfacing

Major gains in productivity arise from reuse of computer-based product data in different application areas. Some applications will access data that is in the CAD/CAM data base, others will need to transfer data in and out of the data base. It is important that facilities exist within the CAD/CAM system to allow a company to interface it to other systems. Another very useful feature of some CAD/CAM systems allows a company to add its own, company-specific functions to the system. These functions are then available to users of the system in the same way as vendor-supplied functions.

9.7.4 Data Management

Initially, great weight is placed on the availability of individual functions within a CAD/CAM system. A little later, the focus switches to the ability of users to learn about the system and to use it on a day-to-day basis. Generally, it is only much later that the importance of good data management capabilities within the system is realized. While there is just a limited amount of data in the system and only fairly simple operations are being carried out on this data, data tends to be managed by its owners on an individual basis. However, as the amount of data within the system and the number of users increase, it becomes important to have available efficient methods of data enquiry and access. Similarly, in time, an increasing amount of product data within the system is modified, or used in other products, and in the absence of good data management it becomes more difficult to maintain knowledge of the data status and relationships.

9.7.5 Future System Development

It is only too easy at system selection time to concentrate too much on short-term issues concerning initial system use. It should not be forgotten however that the system will probably be in use by the company for at least ten years, and that product data built up using the system may need to be accessed 20 or 30 years later. The need for a system that works well in the short term must be balanced with the requirement for a system that will be maintained and developed in the long term.

9.7.6 User Interface

It is important that the users feel at home when working with the system. The user interface (made up of the menus, messages, commands, and so on), allows the user to work with the system. It should be easy to learn, easy to understand, and easy to use. A poor user interface can annoy users to the point at which they refuse to work with the system.

9.8 CAD/CAM SYSTEM SUPPORT

The CAD/CAM manager and the CAD/CAM team have crucial roles to play in the development of the use of CAD/CAM in the company.

9.8.1 The CAD/CAM Manager

The CAD/CAM manager will be in a difficult position, under pressure from top management (wanting to ensure that CAD/CAM produces the expected productivity gains), from middle management (who will be anxious about CAD/CAM causing them to fall behind project schedules), and from users (complaining about everyday problems of system use and operation). The CAD/CAM manager must not spend too much time in responding to this pressure, since there are more important positive actions to be taken in respect of CAD/CAM. For example, it is necessary to ensure that the system runs as efficiently as possible on a day-to-day basis, that new users and projects are attracted to CAD/CAM, that suitable training is available and given throughout the company, that plans are made for future system use, and that other systems in the company are interfaced to the CAD/CAM system.

Top and middle management must give full support to the CAD/CAM manager on an ongoing basis. If it is felt that the CAD/CAM manager really is not doing the job in the correct way, then a clean break is preferable to a long, drawnout, backstabbing approach that can only be harmful to the image and use of CAD/CAM within the company.

In view of the many different tasks involved, the CAD/CAM manager's position within the organization must be well defined. This position must carry with it enough authority to implement decisions taken to develop CAD/CAM use throughout the company.

The CAD/CAM manager must be a good organizer and a good communicator, and must be able to present a good image of the company. The CAD/CAM manager will be in contact not only with all levels of people within the company, but also with the system vendor, with visitors interested in seeing CAD/CAM at work, and with potential clients of the company.

If possible, the person who is to become CAD/CAM manager should be involved in the initial stages of learning about CAD/CAM and selecting a system. Early and deep involvement in those processes will increase the desire of the CAD/CAM manager to ensure that CAD/CAM is successfully introduced and used.

The CAD/CAM manager must be given enough resources, in particular support staff, to make certain that all the tasks associated with successful implementation of CAD/CAM can be carried out.

9.8.2 The CAD/CAM Team

The size of the CAD/CAM team will vary from company to company. However, there seems to be a fairly widespread habit in companies of all sizes of trying to implement CAD/CAM without enough support staff. Typically this leads to failure to attain the true potential of CAD/CAM, and to overwork and disillusionment of the CAD/CAM team members. Team members are attracted to other more rewarding positions and the CAD/CAM team becomes even smaller.

Perhaps part of the fault lies with those responsible for initially cost justifying the system. If forced to show that the system will be as productive as possible, they may well cut back on overall CAD/CAM costs by reducing the number of staff involved. Afterwards, of course, it is difficult to justify hiring more personnel.

The CAD/CAM team must have a good mix of engineering and EDP knowledge. It would be a mistake to staff the team with only EDP specialists, as they would have little success in activities such as training and understanding user requirements. Conversely, some EDP knowledge is needed to operate the system efficiently and to interface it to other activities in the company.

9.9 CAD/CAM SYSTEM DEVELOPMENT

Ideally, CAD/CAM system development should occur both through the implementation of new versions released by the vendor and from internal developments linking it to other systems in the company.

9.9.1 Vendor-Supplied Developments

Great advances have been made in improving the capabilities of CAD/CAM systems over the last 20 years, and it can be expected that important advances will also be made in the future.

A company should ensure that it selects a system with a potential for future development. With a few exceptions, it appears that individual CAD/CAM systems have a limited lifetime (of about 10 years) after which it apparently becomes very difficult to maintain them at the level of advancing technology. Among new technologies which have caused problems to the smooth development of CAD/CAM systems are the change from

16 bit to 32 bit computers, the change from vector to color raster graphic screens, the introduction of surface and solid modelling, implementation of stand-alone engineering workstations, and new software, data management, and numerical analysis techniques.

During the initial selection process it is important to evaluate the development history and potential of a system and its vendor. By looking at the developments made to a system over the last few years, it is possible to learn something about the ability of the vendor to develop the system. Existing users of a system can be asked about the relationship between a vendor's development plans and the actual developments released. It is also useful to look at the development plan proposed for the coming years. However, this information may be of little use if the vendor has failed to meet targets in the past.

9.9.2 Company-Specific Developments

Developments to the CAD/CAM system made by the company generally fall into one of two categories. The first category includes the addition of specific functions not available within the system. However, as more and more functions become available within CAD/CAM systems there is much less need for small- to medium-sized companies to develop their own functions. The second category of developments is mainly concerned with interfacing the CAD/CAM system to other systems in the company. Improvements in productivity result from the reuse of computer-based product data. They will not occur if the CAD/CAM system is seen as a closed system into which all data is entered by hand, and from which data can only leave in the form of drawings. Most companies have a need to interface other systems to the CAD/CAM system. Interfaces invariably involve data transfer, and problems may well arise with vendors loathe to disclose where and how data is stored in their systems. Company developed interfaces must be well documented. All too often they are not, and knowledge of how they work remains in the head of one person. Such interfaces are very difficult to operate and maintain.

9.10 CAD/CAM SYSTEM OPERATIONS

To achieve maximum benefits from the use of CAD/CAM, the system should be in productive use as much as possible. This requires good planning and organization. System hardware and software must be in working order and available, and all other resources (e.g., tapes, paper) must be in place. The system should be backed up frequently to prevent unnecessary loss of working data. Data archiving procedures must be implemented. Maintenance must be scheduled to be carried out without disturbing productive system use. Workstation scheduling must be implemented to ensure that users can plan their work in advance, and have access to the system when they need it. To avoid undue disturbance to system users, the system should be modified as rarely as possible, and before any modifications are put in place they should be fully tested. Efficient operations, data management, and project management procedures should be put in place as early as possible.

9.11 ONGOING CAD/CAM MANAGEMENT

It is not enough to select the correct CAD/CAM system and install it properly. The real benefits of CAD/CAM appear in the long term. They will only be achieved if ongoing use and development of the system is correctly managed.

9.11.1 Project Management

Without CAD/CAM, projects to develop and modify the company's products require good project management. The use of CAD/CAM in no way reduces the need for project management. On the contrary, because it introduces new resources and reduces overall time cycles, CAD/CAM actually increases the need for good project management.

It is particularly important at early stages of CAD/CAM use (when managers and users are unfamiliar with use of the system) that close attention be paid to the progress of projects using CAD/CAM.

9.11.2 Company CAD/CAM Documentation

The CAD/CAM system vendor will supply some documentation describing use of the system. Often this documentation does not describe the exact configuration of the system that the company is using, and it cannot describe the way that the company wants to integrate use of the system with everyday operations. However, users of the system need to have available up-to-date and complete information describing utilization of the system within the company environment. This information should be available in a company CAD/CAM manual (perhaps computer based) in which users can rapidly access required information. Such a manual could include, for example, a description of the standard company-specific parts that have been stored in a parts library. In the absence of such information it can be seen that a loss of productivity will occur either because the user will have to look for the information somewhere else (but where?) or because the user will design a new part that may be exactly the same as an existing one. The company CAD/CAM manual should contain all necessary information, but should not be so thorough that it becomes unusable or overly restricts the creativity of users.

9.11.3 Data Library

One of the major sources of increased productivity due to the use of CAD/CAM derives from the reuse of computer-based product data. Two of the many ways to benefit from reuse of this data occur when the same data is used in manufacturing engineering as in design engineering, and when an existing product is slightly modified. One obvious prerequisite for reuse of product data is that it can be stored in the first place. Another, less obvious but also very important prerequisite is that the stored data be readily accessible. Without proactive management intervention, data has an unfortunate habit of becoming unfindable, inaccessible, in the wrong format, and generally useless. From the earliest stages of CAD/CAM use, attention must be paid to building up the store of readily retrievable and reusable data.

9.11.4 System Monitoring

Unless the use of the CAD/CAM system and the use of CAD/CAM techniques are closely monitored, there is no way of knowing whether they are productive or whether they come up to original targets. Similarly, unless information is available on current use it is not possible to plan for further use. Many CAD/CAM systems can automatically provide some information on their use, although in some cases the information is very limited in both quantity and usefulness. As for monitoring the effectiveness of the use of CAD/CAM techniques, each company must define the factors that it needs to measure and the methods by which they will be measured. Both methods and factors should be defined before initial use of the system.

9.11.5 Forward Planning

It must be remembered that introduction of CAD/CAM is not a one-step solution to a company's problems. Correct choice and effective implementation of a CAD/CAM system is only the first stage in a company's use of CAD/CAM. Initial use of CAD/CAM may cause a drop in productivity, with benefits only coming in the long term. As time goes on, use of CAD/CAM will spread throughout the company. There will be more users of CAD/CAM, CAD/CAM will be used for new applications, CAD/CAM systems will become more powerful, and other systems in the company will be interfaced to the CAD/CAM system. An ongoing plan, developed from the long-term strategy, must be put in place to ensure efficient and expanding use of CAD/CAM.

9.11.6 Learning from Others

During the early stages of CAD/CAM system selection, implementation, and use, a company will learn very quickly about CAD/CAM techniques. However, there is always a lot more that can be learned, in particular from other companies. One way to do this is to attend user group meetings and CAD/CAM conferences. Another way is to visit other users situated nearby (or using the same system, or making similar types of product), and trying to learn from the way in which they use CAD/CAM.

9.11.7 Doing It Together

CAD/CAM is a technology that can lead to productivity increases through better use of information. Information is a company-wide resource. For CAD/CAM to be successful, top management and managers and users from different departments must work together.

FURTHER READING

There are many hundred publications on the subject of CAD/CAM, but rather than give a long and incomplete list the following have been selected:

As a general introduction to CAD/CAM:

John K. Krouse. *What Every Engineer Should Know About Computer-Aided Design and Computer-Aided Manufacturing.* Marcel Dekker Inc., New York, 1982.

Making a Start in CAD, All You Wanted to Hear About the Subject But Didn't Know Who to Ask. IPC Business Press Ltd., Sutton, England, 1982.

As an aid to understanding the overall engineering and manufacturing operations of a company as a background to the implementation of computer systems such as CAD/CAM:

Joseph Harrington, Jr. *Understanding the Manufacturing Process.* Marcel Dekker Inc., New York, 1984.

As a general introduction to geometry modelling in CAD/CAM:

I. D. Faux and M. J. Pratt. *Computational Geometry for Design and Manufacture.* Ellis Horwood Ltd., Chichester, 1979.

A. Bowyer and J. R. Woodwark. *A Programmer's Geometry.* Butterworths, London, 1983.

Computers in Industry. Volume 3, Numbers 1 and 2, North Holland, Amsterdam, 1982.

As an introduction to using a CAD/CAM system:

J. C. Lange. *Interactive Computer Graphics Applied to Mechanical Drafting and Design.* John Wiley & Sons Inc., New York, 1984.

As an introduction to numerical control systems and programming:

P. Bézier. *Emploi des machines à commande numérique.* Masson et Cie, Paris, 1970.

J. Pusztai and M. Sava. *Computer Numerical Control.* Reston Publishing Company Inc., Reston, Virginia, 1983.

As an introduction to group technology:

C. G. Gallagher and W. A. Knight. *Group Technology.* Butterworths, London, 1983.

On the psychological and sociological impact of CAD/CAM:

J. P. Poitou. Les consequences psychologiques et sociologiques d'une nouvelle forme d'informatisation: Les techniques de conception assistée par ordinateur C.A.O. University of Provence, Aix, 1983.

GLOSSARY

To avoid breaking up the flow of the text, the definitions of words used in the book have been grouped in this glossary. Many of the words are part of CAD/CAM jargon and are used with different meanings by different people. The glossary is not meant to be a study in etymology. It is only intended to give some help to design engineers and manufacturing engineers with little knowledge of computers, to data processing specialists with little knowledge of engineering, and to managers with little knowledge of either engineering or computers.

Acceptance test A test used to check that the actual performance and capabilities of a delivered system, hardware, and software are in agreement with the promised specifications.

Access methods Data handling techniques used in particular for transferring data between main memory and peripheral devices.

Access time The time to access data (i.e., the time between the moment at which data is called for and the moment at which it becomes available).

Accuracy A measure of how the actual position of a device corresponds to the requested position.

Address A unique label that identifies a particular memory location.

Addressable points The points on a device that may be specified by absolute values.

AFNOR (Association Francaise de Normalisation) French standards organization.

Algorithm A well-defined procedure for solving a problem.

Aliasing A technique used to minimize the staircase effect seen when displaying diagonal lines on a raster device.

Alphanumeric display Device consisting of a typewriter-style keyboard and a cathode ray tube (CRT) screen capable of displaying alphanumeric text (but not graphics).

Alphanumeric text Text made up only of alphabetic and numeric characters and a few special characters (such as %,?).

AMT (Advanced Manufacturing Technologies) Manufacturing technologies (e.g., robotics, FMS, DNC) in which computer systems play an important role.

Analog Continuous-value representation of a physical quantity (as opposed to digital).

Annotation Addition of text to a drawing or geometric model.

Application One of the sets of tasks to be carried out in the overall product development and production process (e.g., design, machining, quality control).

Application program (1) A program written for a specific application (e.g., meshing), (2) A user-written program (as opposed to a system program normally written and supplied by a computer manufacturer or a computer systems manufacturer).

APT (Automatically Programmed Tools) The earliest and one of the most frequently used part-programming languages. Output from APT describes the cutter tool path.

Archive To store data on a storage medium (generally a magnetic tape) for long-term reference purposes.

Array An arrangement of elements in one or more dimensions.

AI (Artificial Intelligence) The capability of a computer system to perform functions analogous to human abilities of learning, reasoning, decision making, and self-improvement.

ASCII (American Standard Code for Information Interchange) An 8-bit standard code used to encode keyboard characters and some special purpose symbols.

Assembly A number of basic parts or subassemblies, joined together to perform a specific function.

Assembly language A low-level computer language in which individual instructions, often written in mnemonic form, correspond closely to machine code instructions.

Assembly drawing A drawing representing, in their correct positional relationships, a group of parts constituting a major subdivision of the final product.

Associative dimensioning system A system allowing automatic updating of the dimensions of entities to which they are linked as the entities are changed.

Associativity A logical link between entities and/or attributes in the CAD/CAM data base.

Attribute A characteristic associated with an entity or data item in the CAD/CAM data base.

Automated process planning Use of a computer program to generate the process plan for a new part on the basis of known plans for existing, similar parts.

Automatic dimensioning Automatic measurement of distances and placing of extension lines, dimension lines, and arrowheads.

Automation Mechanical or electronic systems that eliminate or reduce the need for manual processing.

Auxiliary storage Peripheral storage devices such as disks, tapes, and floppy disks.

Axonometric projection A projection in which a drawing of a three-dimensional object has all lines to exact scale and appears distorted. Isometric, dimetric, and trimetric projections are special cases of axonometric projection.

Backup To copy data to keep as a reference in case the original data is lost.

Ball-end mill A milling machine cutter shaped so that the envelope of the cutting edges is a hemisphere at the end of a cylinder.

BASIC An easy-to-use medium-level computer language often used to solve engineering problems.

Batch (1) A group of identical parts moving through the manufacturing cycle. (2) A computer program that once fed to the computer, is processed without further human intervention.

Baud A unit of signalling speed used as a measure of data flow between computers and/or communication devices. In the case of binary signals, one baud equals one bit per second.

Benchmark A set of standards used in testing software, hardware, or a system. Benchmark tests are carried out before purchase.

Bezier representation A mathematical representation used in geometry modelling of complex curves and surfaces.

Bicubic representation A mathematical representation used in geometry modelling of complex surfaces.

Binary Referring to the base 2 number system.

Bit An abbreviated form of "binary digit." The smallest unit of information in a binary-based system. A bit can be set equal to 0 or 1.

Bit map Memory used to store information on the attributes of each pixel on a raster screen.

Black box Description of a system about which only inputs and outputs are known, but not the way in which internal processing takes place.

Boolean algebra An algebra that defines the rules for manipulating variables with symbolic logic.

BOM (Bill of materials) A list of all the subassemblies and parts that make up an assembled product. Their material and quantity are also defined.

BPI (Bits per inch) Number of bits of binary data that are stored on a one-inch length of magnetic tape.

BS (British standard) British industrial standard.

B-spline representation A mathematical representation used in geometry modelling of complex curves and surfaces.

Bug An error in a computer program or system.

Byte Eight adjacent bits treated as a basic unit of information and handled as a unit. In ASCII code, one alphanumeric character can be stored in one byte.

CAD (Computer-aided design) A computer-based tool, using interactive graphics techniques, that is used in translating a requirement or concept into an engineering design, the geometry of which is stored as a model in a computerized data base.

CAD/CAM (Computer-aided design engineering; computer-aided manufacturing engineering) A computer-aided technique for improving the efficiency of design engineering and manufacturing engineering activities. Distinguishing features are the use of interactive graphics, geometry modelling, and reuse of part data in several activities.

CAE (Computer-aided engineering) All computer-based techniques used in the design engineering and manufacturing engineering areas.

CAM (Computer-aided manufacturing) The use of a computerized data base from which information can be extracted and used as direct input to control manufacturing equipment.

CAM-I (Computer-Aided Manufacturing-International) A multinational not-for-profit organization furthering R&D in the CAD/CAM area.

Canonical form A standard numerical representation of data.

CAP (Computer-aided production) The use of computers in the production process.

Capacity The amount of information that a memory can store.

CAPP (Computer-aided process planning) See Automated process planning.

CAQA (Computer-aided quality assurance) The use of a computer to improve the manufacture, useful life, and reliability of a product.

Cassette A peripheral data storage device. A small cartridge containing two reels between which magnetic tape transits.

CAT (Computer-aided testing) (1) The use of a computer for real-time monitoring of real-life tests performed on a part. (2) The use of a computer to test out a computer-based design by simulation programs.

Cell A grouping of manufacturing tools and part-handling devices capable of carrying out, without human intervention, a number of manufacturing operations.

Channel A communication path.

Change order A formal notice that a specification (e.g., an engineering drawing) has been changed.

Character An alphanumeric, numeric, or special graphic symbol.

Character font The style of a character (e.g., gothic).

CIM (Computer-integrated manufacturing) The integrated computerization of all functions of a manufacturing company.

CL file (cutter location file) A data file containing the definition of the path of the center of a tool as it machines a part.

Clipping The process of removing or "clipping off" the part of an overall picture that is outside a defined boundary of a displayed image.

CNC (Computer numerical control) A numerical control system in which a small computer dedicated to a tool is used to control the tool. The computer contains memory, in which the part program is stored, and executes the program block by block.

Code A system of symbols and characters and rules for their interpretation. Such systems are used in computing to represent instructions and data.

COM (Computer output microfilm) Microfilm containing an image of data generated by a computer.

Communications network A system to move information between a number of computers and their associated peripherals.

Compact II A widely used part-programming language.

Compatibility The ability of software, data, or hardware to interface with other software, data, or hardware.

Compiler A program that translates high-level programming language source instructions into machine code.

Complex curve Complex curves and surfaces are those that cannot be mathematically represented by analytic equations. They are often represented by polynomials.

Computer Electronic equipment capable of solving problems by accepting data, performing requested operations on the data, and outputting the results of these operations.

Computer-dependent program A program that will only run on one specific computer. It would require modification to work on another type of computer.

Computer graphics The methods and techniques associated with computer processing, input, or output of graphical data representations.

Conceptual design Design leading to a general, undetailed notion of a product in agreement with specifications.

Configuration The detailed combination of computer and peripheral devices making up a computer system or installation.

Console terminal A terminal used for input of system commands and output of system messages.

Conversational mode A mode of communication between a computer and a user at a terminal in which each entry at the terminal elicits a response from the computer.

Core A type of main memory.

CPU (Central processing unit) The part of a computer that controls the interpretation and execution of program instructions.

Crash Failure of hardware or software causing involuntary termination.

Crosshairs Two perpendicular lines, the intersection of which is used to select a point on a target.

Crosshatching An engineering drawing technique of shading with intersecting sets of parallel lines.

CRT (Cathode ray tube) A vacuum tube in which an electron beam is projected onto a fluorescent screen to produce a luminous spot. The beam can be varied in position and intensity, thus producing a visible pattern.

Cubic spline A third-order polynomial curve often used in the representation of complex curves.

Cursor A visible moveable marker (often crosshairs or an underscore) used to indicate the location of the next point on a graphics screen or on a digitizer at which an action is to take place.

Customization Modification of a general-purpose system to meet a particular user's requirements.

Cut (1) To intersect a 3D object with a plane to derive sectional data. (2) To remove material from a workpiece.

Cutter The tool that cuts chips from a workpiece.

Cutter offset The distance from the part surface to the center of the cutter.

Cutter path The path of a tool cutting a part.

Cutting tool A tool that cuts chips from a workpiece.

Data A representation of information in a formalized structure.

Data exchange The transfer of data from a CAD/CAM system to another system.

Data management The management of data produced by CAD/CAM system users.

Data redundancy Repetition of the same information in the data base.

Database A collection of organized, interrelated data with corresponding access, protection, and communications systems.

Dedicated computer A computer which works full time carrying out just one function.

Detail drawing A drawing of a single part of an assembly, containing all the information needed to define the part.

Developable surface A surface which can be unrolled onto a plane without deformation.

Device independent Data or software that is not limited to use on just one physical device supplied by a particular vendor.

Digital A discrete representation of a physical quantity.

Digitize To convert an object, drawing, or picture into information in digital form.

Digitizer An input device which converts physical location coordinate information into digital data readable by a computer.

Dimension A numerical statement of the distance or angle between two features on a part.

Dimensioned drawing A part drawing including relevant dimensions and texts.

DIN (Deutsches Institut für Normung) German industrial standards organization.

Direct access An access method in which data is accessed directly without accessing preceding items.

Discrete parts manufacturing Manufacturing of discrete, different parts as opposed to bulk or batch manufacturing.

Disk A peripheral memory device consisting of one or a stack of rapidly rotating thin metallic disks (also called platters) with magnetic surfaces on which information can be stored.

Diskette A peripheral memory device. A small, easily transportable version of a disk.

Display A data output device capable of giving a visual, pictorial representation of data.

Display controller An electronic device that interfaces information for graphics display generated by a computer with a graphics display device.

Display parameters Data which control the appearance of graphic entities (e.g., line font).

Distributed processing A computer network containing several interconnected processors in each of which data processing tasks relating to local needs can be performed (as opposed to centralized processing or mainframe processing).

Distribution The process of transferring an assembled product to the customer.

DNC (Direct numerical control) A system connecting a set of numerically controlled machines to a common memory for part program storage. Provision is made for online control of the machines, and collection, display, and editing of data and programs related to the numerically controlled process.

Document A physical medium and the data recorded on it.

Double precision The use of double the usual number of words to store a numeric value, thus increasing the precision of the stored value.

Downtime Time during which a computer or system is unavailable for productive use.

DP (Data processing) See EDP.

Drafting The production of detailed drawings.

Drawing A representation by lines of an object on paper or a similar medium.

Drive The peripheral unit that reads and writes data on a mass storage device.

Drum plotter A plotter on which the paper moves over a cylinder or drum. The pen moves horizontally along the top of the drum. Plotting in the perpendicular direction is carried out by rotating the drum, and thus the attached paper.

Dumb terminal A graphics terminal with no local ability to perform graphics processing.

Edit To modify the content or format of data.

EDM (Electrodischarge machining) Removal of material by electrolytic methods.

EDP (Electronic data processing) The broad field of electronic computers and their treatment of data by programs.

Elapsed time The real-world time between the moment at which a user initiates an action, and the moment at which the user sees that the computer has completed the action. When a computer handles several tasks at the same time, the actual processing time for a particular task will be less than its elapsed time.

Electrostatic plotter A device providing graphic and alphanumeric output by electrostatic printing techniques.

Element A low-level item in a structure.

End mill A milling machine cutter with teeth on both the end and the side of the tool. It may be operated with its axis perpendicular, parallel, or at an angle to the work surface.

End user The person who uses a computer-based system in everyday work (but does not develop the system).

Engineering design Functionally complete design, yet lacking some details.

Engineering workstation A single-user system made up of one CPU, one graphics workstation, one keyboard, and perhaps some other peripheral devices. The system is made available by a single vendor, and is often packaged in and on a single cabinet.

Entity A basic building block used in representing a part (e.g., an arc, a character, a patch of a surface, a cube).

Error message A message issued by the system to inform a user that an error has occurred in data input or program execution.

ESPRIT (European Strategic Project for Research and Development in Information Technology)   A joint effort by the European Economic Community and industrial companies based in Europe to develop the European information technology industry.

Evaluation   An investigation to see how a particular system or package can be used in a given environment.

EXAPT (Extended subset of APT)   A part-programming language.

Expert system   A computer program that uses a knowledge base and inference procedures to solve a problem in a particular field.

Exploded view   A view of an assembly in which the individual parts are clearly separated.

Fabrication   Within the overall manufacturing process, production operations as opposed to assembly operations.

Face mill   A milling machine cutter operated with its axis perpendicular to the work surface.

Factory automation   An all-inclusive name covering all the automated and computer-assisted techniques applied in physically producing the product.

Family of parts   A collection of different parts with a "family" resemblance that allows them to be treated in similar fashion.

FEA (Finite-element analysis)   Engineering method used in determining by computer how a part will behave. For the purpose of the analysis, the part is divided up into a large (but finite) number of small elements.

Feed motion   The relative motion between successive cutting actions of a tool on a workpiece.

Feed rate   The speed of a cutting tool relative to the workpiece in the direction of the feed motion.

File   A uniquely named storage area (and its contents) on a peripheral data storage device.

Finite-element mesh   A collection of small elements representing a part.

Finite-element method Treatment by the finite-element method involves three phases: generating the finite-element model (pre-processing), finite-element analysis, and output of results (post-processing).

Firmware Software implemented in Read Only Memory.

Fitting Calculation of the "best" curve or surface to represent a given set of points.

Fixture A device used to locate and hold a workpiece.

Flatbed plotter A plotter that draws on a medium (e.g., paper) mounted on a flat table.

Flicker Image intermittence, or brightness variation, on a vector refresh display, occurring when the total time needed to display a large number of vectors exceeds the decay time of the phosphor.

Floppy disk A peripheral data storage medium. A small, flexible disk that rotates inside a paper sleeve.

FMS (Flexible Manufacturing System) A combination of DNC machine tools and automated material handling systems that can be used to produce a variety of parts.

Fondue Swiss cheese-based dish.

Font A given size and style of character representation.

Forecast Predict sales for a long-term planning horizon.

Format The physical arrangement of data on an input or output device.

Forming tool The tool that deforms a workpiece to give it a required shape (e.g., a die).

FORTRAN (Formula translation) A high-level programming language widely used to handle engineering and other technical problems.

Function generator Special electronics in a graphics terminal to carry out certain frequently used actions (e.g., character generation).

Function keys Specific keys used with a terminal. A predefined function is invoked when a given key is depressed.

Geometry model A computer-based mathematical representation of the shape of a part.

Geometry modeller Ideally, the geometric modeller, which is used to define the model of parts, should be able to model any part, calculate all geometric values relative to the part, produce a model that is usable by all applications, and offer a good response time.

Graphics The display of data by the use of nonalphanumeric means (e.g., by using lines, points, colors, etc.).

Graphics screen A CRT used for the display of graphics data.

Graphics tablet A flat surface by which coordinate pairs identified by a cursor or stylus can be transmitted to a computer.

Group technology Grouping of parts with similarities in design and/or manufacturing into characteristic families of parts to facilitate treatment of these parts and new parts which belong to these families.

Hand tool A working instrument used by hand (not machine driven).

Hands-on Training involving the actual operation of the system.

Hardcopy Direct copy to a plotter (without program intervention) of the image displayed on a graphics screen.

Hardware The physical components of a computer system (electronics, mechanical parts, etc.) as opposed to the software.

Hardware independent Describes data or a program that may be used on more than one type of hardware (i.e., it does not require special features only available with one type of hardware).

Help message A message transmitted by a computer system to help or assist a user in interpreting a result, or in deciding what to do next.

Hidden lines Lines representing the edges or surfaces of a three-dimensional object that would not be visible from a given viewpoint.

High-level language A programming language in which each instruction generates several machine code instructions. It is generally English-like with the instruction's name indicating its function.

Host computer The primary computer in a computer system or network. It provides services such as number crunching and data base management which other computers and processors in the network are not configured to carry out so efficiently.

IGES (Initial graphic exchange specification) A preliminary but currently incomplete standard defining how data should be transferred between CAD/CAM systems.

Information A valuable resource.

Inquiry A request for information stored in the system.

Inspection Examination of a product to determine its quality relative to predetermined specifications.

Integrated system A system capable of carrying out several applications (e.g., design and manufacturing applications).

Intelligent terminal A terminal with its own CPU for local processing.

Interactive graphics The capability to perform graphics operations on a display with a very quick response from the computer to each operation.

Interface The link allowing two independent systems to work together.

Interference A clash or collision between two parts.

Interpolation The insertion of intermediate information based on some assumption of the way in which values change between known points.

Inventory A list of stock.

I/O device (Input/output device) Device used to communicate data and commands from/to the user or a machine to/from the computer.

IPS (Inches per second) A measure of the number of inches of magnetic tape processed by a magnetic tape unit in one second.

Island of automation A part of the overall manufacturing process that has been automated separately from other parts.

Isometric view A view having the projection plane equally inclined to three perpendicular axes.

Jig A device used to locate and hold a workpiece, and guide and control the tool.

Job In DP terminology, a task, a program.

Joystick A data-entry device used to displace a cursor to a suitable position, and specify the coordinates of this position.

KB (Kilo byte) A measure of memory. 1 kb = 1024 bytes.

Keyboard A data-input device. Most keyboards used in CAD/CAM have standard "QWERTY" typewriter-like keys, some additional function keys, and a device for displacing a cursor.

Kinematic analysis Analysis of the movement of parts or mechanisms.

Knowledge base The known facts and rules in a particular field used by an expert system.

Lathe A machine tool for turning and shaping parts.

Layer A logical concept in CAD/CAM used to distinguish different groups of data within a given drawing. Layers may be thought of as a series of transparent sheets that may be overlaid in any order, but have no depth. The user can select which layers are to be displayed at any one time.

Layout drawing A drawing, often to scale, made during the design process to help the designer develop a workable design. It may omit unnecessary details, which will however be shown in detail drawings.

Learning curve (1) A graph showing on the vertical axis the increase in percentage productivity a new user achieves with a system. The horizontal axis shows the length of time the system has been used for. (2) A graph showing on the vertical axis the increase in overall productivity of a company due to use of a system. The horizontal axis shows the time since system installation.

Learning time The time that a new user or operator takes to learn to use a system or machine as efficiently as an experienced user or operator.

Library (1) A collection of programs. (2) A collection of pre-defined parts or subassemblies stored in a CAD/CAM data base and available for use by all users, thus obviating the need for redesign.

Light pen An optoelectronic device in a pen-like holder that is used to select or position an item on a graphics screen.

Line font The pattern for the appearance of a curve on a graphics display.

Line printer A printing device that can print an entire line of characters at a time.

Linear interpolation A technique that reduces the number of data points that have to be stored by assuming that intermediary points lie on the straight line connecting two consecutive points.

Listing A computer-produced, readable list on paper.

Machine code The code understood directly by a computer and translated directly into electronic circuit operations.

Machine tool A powered, permanently installed tool.

Machining center A numerically controlled multifunction machine tool equipped with an automatic tool changer and/or pallet changer.

Macro An often-used, named collection of many commands that can be instigated by referring to the macro name, thus avoiding unnecessary repetition.

Magnetic tape A data storage medium, suitable for long-term storage owing to its relatively low cost and easy accessibility.

Main storage Main memory, the internal memory of a computer as opposed to peripheral memory devices.

Mainframe A powerful, central computer capable of carrying out large numbers of calculations, managing data bases and supporting large numbers of terminals. These computers generally have a word length of at least 32 bits.

Maintenance An activity intended to ensure that systems stay in satisfactory working condition.

Maintenance cost The cost of vendor- or third-party-supplied maintenance.

Manufacturing Shop floor-related operations in the overall product process.

Manufacturing control system The system that controls shopfloor activities such as process automation, in-process quality testing, material and machine monitoring, etc.

Manufacturing planning system The system used to plan availability of materials, machines, and people in order to manufacture products for the required date.

Manuscript A form used by a part programmer for the detailed manual listing of part program instructions.

Mass properties Physical engineering information about a part (e.g., area, volume, center of gravity).

Master production scheduling Use of a long-term forecast to determine long-term production requirements.

Materials handling The movement of materials from one location to another without modification of their form or nature.

Matrix A two-dimensional rectangular array of quantities.

MB (Mega byte) A measure of memory. 1 MB corresponds to 1048576 bytes of memory.

MDI (Manual data input) Manual insertion of data into a control unit (e.g., by keyboard) as opposed to input by tape or communications line.

Mechanism Part of the workings of a machine.

Memory That part of a computer system that is used for storing data.

Memory plane Memory used to store information on the attributes of pixels on a raster screen.

Menu A list of options or functions available to the user. Menus are sometimes drawn out on overlayable sheets that can be placed on graphics tablets. Alternatively they can be displayed on a graphics screen.

Mesh generation Technique used to generate a large number of small elements corresponding to a larger element. The subelements can then be handled by finite-element analysis techniques.

Metal cutting The process of converting raw metal into a finished part by cutting away unwanted metal.

Metal forming The process of converting raw metal into a finished part by deformation. Little material is lost.

Microcomputer Generally, a computer whose CPU is a single integrated circuit and which has limited memory.

Microfilm Film on which drawings are photographed in reduced size for convenient storage.

Microsecond One millionth of a second.

Milling machine A machine tool for cutting parts with a rotating milling cutter.

Millisecond One thousandth of a second.

Minicomputer Generally, a computer with a 16-bit word length and limited memory addressing capability.

MIPS (Millions of instructions per second) A measure of a CPU's computing power.

MIS (Management information system) The part of a company's EDP system that relates to business activities.

Model (1) A geometrically accurate and complete mathematical representation of a part as stored in a CAD/CAM system. (2) An early version of a part, often to scale, built to test certain features.

MODEM (MOdulator-DEModulator) The device that allows a terminal to be connected by a communications line to a computer.

Mouse A data input device, used to define relative position. It does not need to be placed on a particular device such as a graphics tablet, but works on any flat surface. It is often used to control the relative position of a cursor on a screen.

MRP (Material requirements planning) Based on stocks, work-in-progress and orders, an MRP system plans the order of raw materials and purchased parts.

MRP 2 (Manufacturing resource planning) An extended MRP system concerned with allocation and management of machines, manpower, and materials as a function of manufacturing requirements.

NC (Numerical control) The use of coded numeric information, rather than an operator, to control machine tools.

NDT (Nondestructive testing) Use of techniques to check the physical limits of parts without breaking them.

Nest To arrange many part shapes on a large sheet with the intention of minimizing the amount of material wasted when the parts are cut out.

Network A system of communications lines and devices.

Node (1) An individual processor in a communications network. (2) Within a mesh, a point of intersection between two mesh lines.

Number crunching Computers or programs used mainly for carrying out calculations and not for data base management, user interaction, or text handling.

OA (Office automation) The use of computers for such functions as word processing, electronic mail, filing, and report writing.

Object program The coded output of an assembler or compiler.

OEM (Original equipment manufacturer) A company that buys a computer system from a computer manufacturer, enhances it, and then sells it to an end user.

Offline Equipment or devices in a computer system that are not under the direct control of the computer.

Online Equipment or devices in a computer system that are under the direct control of the computer.

Operating system Software supplied by the computer manufacturer which controls the execution of programs and the data flow to and from peripheral devices.

Optimization Adjustment of a process to obtain the best obtainable set of operating conditions.

Orthographic projection The projection used most often for engineering drawings.

Overlay (1) Masks which can be fitted over function keys to identify their functions. (2) The process of reducing the length in memory of a program by swapping parts of the program that are not used simultaneously into the same section of memory when required.

Page   A single display image on a screen.

Paint   To draw an image on a display screen.

Pallet   A platform on which workpieces are stacked and can be transported.

Pallet changer   An automatic pallet-moving mechanism included in a machining center.

Parameterization   The possibility to give modifiable parameters (rather than fixed dimensions) to a model.

Parametric curve   A frequent representation of curves in geometry modelling.

Part   A physical component, not an assembly of other items.

Part library   A file of often-used parts.

Part number   A number which uniquely identifies a component.

Part program   An ordered set of instructions in a part programming language which, when fed to the control unit of a machine tool, will cause the machine tool to execute the necessary instructions to machine the part.

Part programmer   A person with knowledge of manufacturing, who writes programs in NC languages to control NC machines that will machine parts.

Parts list   The list of all the parts making up an assembly.

Password   A code name given to a user in order to meet security requirements and gain access to the computer.

Patch   In geometry modelling, a small part of the surface of a part.

Pen plotter   A plotter using a pen, rather than electrostatic techniques, to produce a drawing.

Peripheral device   A device distinct from the computer, and used for input, output, and storage of data.

Peripheral mill A milling machine cutter operated with its axis parallel to the work surface.

Perspective view A view of a part giving the same appearance to the eye as the part itself.

Pick To select.

Pixel (Picture Element) A term used to describe the smallest addressable unit of a display surface.

PLC (Programmable logic controller) A microprocessor-based device that can be programmed to control industrial equipment.

Plant layout Arrangement of machines in a plant.

Plotter A computer-controlled device used to make a permanent copy of a screen image on a medium such as paper.

Pocketing A high-level NC programming feature that results in all material being removed from a defined volume.

Postprocessor A computer program that converts output from a computer process into a form acceptable to another process.

Preprocessor A computer program that prepares data for use by a computer program.

Process planning Preparation of the process plan—the list of all machine tool operations which a part must undergo.

Production That part of the overall manufacturing production process that involves direct actions on the workpiece.

Production control The management of the production process.

Productivity A measure of the efficiency of producing a given output from a given input.

Program A set of machine instructions combined to form a useful task.

Programmer A person who writes software to control the actions of a computer.

Programming language A language used to write programs for computers.

Prompt message A message from the computer system requesting the user to carry out a certain action (e.g., to enter coordinate values).

Protocol A set of rules governing use of a computer system, or transmission of data between devices and systems.

Prototype A nearly final model of a product under development, upon which detailed tests will be carried out.

Puck A picking device for data input.

Punched tape Paper or plastic tape into which a pattern of holes in a known code is punched to convey information.

Quality assurance The operation to ensure that manufacturing methods will produce an acceptable product.

Quality control The operation of ensuring that manufactured parts come up to the required standard.

RAM (Random-access memory) Memory that can be both read from and written to directly and independently of memory location.

Random access An access method allowing equal access time to all data independent of their position.

Raster A cartesian coordinate grid that divides up the screen of a display surface.

Raster scan A predetermined, regular pattern of scanning a display device.

Raster screen A graphics screen using raster scan technology.

Redundant data Data that can be removed from a system without losing information content.

Refresh cycle The time between successive raster scans on a raster screen, or between passes through the list of vectors displayed on a vector screen.

Refresh memory That part of memory that holds information on the picture to be displayed on a vector or raster display.

Remote terminal A terminal physically located away from the computer, and connected to it by communication lines.

Repeatability A measure of the closeness of agreement attained when the same operation is carried out several times.

Resolution A measure of the minimum spacing between adjacent distinct points on a machine or screen.

Resource Any part of a computer system.

Response time The time that elapses between the moment that a user requests an action from a system, and the moment that the user receives the response to the request.

RGB (Red green blue) A technique for producing all colors on a display by mixing different proportions of red, green, and blue.

Robot An automated device capable of carrying out some functions normally ascribed to humans.

ROI (Return on investment) A measure of how quickly an investment will pay for itself.

ROM (Read-only memory) Memory whose contents are fixed and under normal operating conditions can only be read, but not modified.

RS-232-C Interface A standard interface for linking peripherals and computers together.

Rubberbanding Interactively changing the shape of a curve on a display while keeping both ends fixed.

Ruled surface In geometry modelling, a surface generated by connecting corresponding points on two space curves by a set of straight lines.

Scalar A quantity possessing only one attribute—magnitude.

Scale The ratio of a displayed image to the actual size of the corresponding object.

Schematics Drawings such as wiring diagrams that do not represent part geometry, but a schematic representation.

Screen The graphics display.

Sculptured surface Another term for a doubly curved or complex surface.

Sequential access A data access method in which all data is read in sequence until the required data is found. The access time depends on the position of the data.

SET (Standard d'échange et de transfert) A French-developed standard format for exchange of data between different CAD/CAM systems.

Shape fill Filling in the inside of a displayed shape's boundaries (e.g., in color).

SFC (Shop floor control) Within the overall MRP2 context, short-term planning, execution, and status reporting of shop-floor operations.

Simulation The process in which a mathematical model of an object is tested in a computer to help predict how the object will behave under given circumstances.

Smart terminal A terminal containing microprocessors capable of carrying out a limited number of functions locally.

Smoothe Reduce fluctuations in data values.

Softcopy Copy, by computer program, the data displayed on a screen to a plotter.

Software Program (as opposed to hardware) instructions controlling the hardware.

Solid modelling A method of geometric modelling in which a complete representation of the exterior and interior of a part is built up.

Source code Computer language instructions, understandable by humans, that are translated by a program into a form in which they can be directly processed by a computer.

Spline In geometric modelling, a polynomial curve passing through given fixed points.

Storage Another term to describe computer memory.

Storage capacity The amount of data that can be stored in the memory of a computer system. Generally it is expressed in units of bytes.

Storage tube A CRT that can retain a visual image for some length of time so that it is not necessary to continually refresh the image.

Stress analysis Analysis of an object by a computer program to investigate how it will behave under a given stress.

Stroke writing Production of an image on a graphics screen by drawing a large number of strokes (or vectors).

Structural analysis Analysis of an object by a computer program to investigate how its structure will behave under different loads.

Styling The process of defining the aesthetic appearance of an object.

Stylus A handheld pen-like object used to provide coordinate information from a tablet.

Subroutine A self-contained set of computer instructions arranged in sequence. It can be called from another subroutine or a program whenever required.

Supercomputer The most powerful type of computer available. Used almost entirely for number crunching applications such as structural and aerodynamic analysis.

Superminicomputer A very powerful version of a minicomputer with a 32 bit word length and virtual memory addressing capabilities. A typical superminicomputer would be slightly less powerful than a mainframe computer, and would be generally used for engineering and manufacturing applications rather than for business applications.

Surface modelling A method of geometric modelling in which the complete external shape of a part is defined.

Surface of revolution A type of surface used in geometry modelling. It is generated by rotating a two-dimensional curve (the generatrix) about an axis (the axis of rotation).

Symbol A representation of an object.

System An organized collection of personnel, hardware, software, and methods capable of accomplishing a set of specific functions.

Tablet A data-input device. A small digitizer, used to input coordinate values to the computer.

Tabulated cylinder In geometric modelling, a surface made by translating a curve along a direction line (the directrix) with upper and lower limits on the distance of translation.

Tape A medium (e.g., magnetic tape, paper tape) for storing data.

Technical publication Publication such as a maintenance document for a product, often containing drawings to aid service engineers or customers to maintain or use the product.

Template A tool or drawing used to define the shape of a part.

Terminal An information entry or exit point in a system or communications network.

Testing Carrying out tests during and at the end of the manufacturing process, to identify deficient products, equipment, and operations.

Three-dimensional That which cannot be completely described by two parameters, but by three.

Time sharing A system in which the CPU of a computer is shared by several users at the same time.

Tolerance The amount by which a part's dimensions may differ from the specified dimensions and still perform or fit properly.

Tool An instrument used to produce a part.

Tool changer An automatic mechanism on a machining center which, upon the appropriate command from the control unit, will remove the tool from the spindle, place it in the tool changer magazine, select the next tool from the magazine and place it in the spindle.

Tool path The locus of a certain part of a tool (e.g., the center of a ball-ended tool) as it machines a part.

Tooling The special tools (excluding machine tools) used in production of a part (e.g., jigs, fixtures).

Topology The connection between parts of a whole, as opposed to their shape.

Trackerball A data input device used to control the position of a cursor.

Turning The working of parts on lathes.

Turnkey system A system that is ready for use at the turn of a key. It is implied that one supplier has total responsibility for hardware, software, and installation.

Two-dimensional That which can be completely described by two parameters.

Two-and-a-half-dimensional That which can be described by varying two parameters while keeping a third parameter fixed. The third parameter may then be changed, but is fixed again while the other two parameters are changed.

Update To modify data or programs to include new or more current information.

Uptime The time during which a system is available for everyday useful work.

Upward compatibility The ability of software developed for one computer in a manufacturer's range to run without modification on later versions of the computer.

User A person using (and not developing) an information system.

User friendly The characteristic of a program that describes its ease of use.

User interface The way in which users see their particular applications as presented by the computer.

VDA (Verband der Automobilindustrie) German automobile industry organization. The VDA interface is a standard format for exchange of data between different CAD/CAM systems.

Vector A quantity possessing two attributes—magnitude and direction.

Vector refresh display A CRT in which the image is refreshed very often (about 30 times a second) and in which the image is produced by the CRT beam drawing straight line vectors between given data points.

Vendor The company that sells a CAD/CAM system to a user company.

Viewing angle The angle at which a geometric model is examined.

Viewpoint The point from which a geometric model is examined.

Virtual storage Memory appearing to a program as real hardware memory, but actually obtained by software methods.

Vision system An AI-based system that automates the recognition of images.

Voice recognition system A data-input device that allows the user to enter commands or data vocally rather than by typing on a keyboard or selecting items from a screen or tablet.

Window A user-selected rectangular area on the display screen.

Wireframe A method of geometric modelling in which a two- or three-dimensional object is represented by its edges.

Word An ordered set of bits in a computer.

Workpiece The piece of material, destined to become a part, on which a tool is working.

Workstation The terminal (including graphics screen, keyboard, etc.) at which a CAD/CAM user works.

WP (Word processing) Use of DP techniques to assist production of texts (e.g., letters, memos) using a screen, keyboard, and printer.

Zone An area of a menu corresponding to a particular function.

Zoom Interactively change the scale of an image on the screen so that all of it, or part of it, appears larger (or smaller) on the screen.

INDEX

ABOUT THE AUTHOR

John Stark is a Director of Coopers and Lybrand Associates Europe, London, England, with particular responsibility for CAD/CAM management consultancy services throughout Europe. He has been involved in the development of CAD/CAM systems and their implementation in the United States and Europe, and has acted as a CAD/CAM expert for the United Nations in the Far East. He is a member of the Society of Automotive Engineers (SAE), American Institute of Aeronautics and Astronautics (AIAA), National Computer Graphics Association, Swiss Computer Graphics Association, as well as the Computer and Automated Systems Association of the Society of Manufacturing Engineers (CASA/SME). Dr. Stark received the B.Sc. (1969) and Ph.D. (1972) degrees from the Imperial College of Science and Technology, University of London, England.